Adult Math Workbook

Basic Math Exercises, Tutorials and Multiple Choice Strategies

COMPLETE
Test Preparation Inc.
WWW.TEST-PREPARATION.CA

Copyright © 2025 by Complete Test Preparation Inc. ALL RIGHTS RESERVED.

No part of this book may be reproduced or transferred in any form or by any means, graphic, electronic, or mechanical, including photocopying, recording, web distribution, taping, or by any information storage retrieval system, without the written permission of the author.

Notice: Complete Test Preparation Inc. makes every reasonable effort to obtain from reliable sources accurate, complete, and timely information about the tests covered in this book. Nevertheless, changes can be made in the tests or the administration of the tests at any time and Complete Test Preparation Inc. makes no representation or warranty, either expressed or implied as to the accuracy, timeliness, or completeness of the information contained in this book. Complete Test Preparation Inc. make no representations or warranties of any kind, express or implied, about the completeness, accuracy, reliability, suitability or availability with respect to the information contained in this document for any purpose. Any reliance you place on such information is therefore strictly at your own risk.

The author(s) shall not be liable for any loss incurred as a consequence of the use and application, directly or indirectly, of any information presented in this work. Sold with the understanding, the author is not engaged in rendering professional services or advice. If advice or expert assistance is required, the services of a competent professional should be sought.

We strongly recommend that students check with exam providers for up-to-date information regarding test content.

Published by
Complete Test Preparation Inc.
Victoria BC Canada

Visit us on the web at
https://www.test-preparation.ca
Printed in the USA

ISBN-13: 978-1-77245-553-3

Version 9 August 2025

About Complete Test Preparation

Why Us?
The Complete Test Preparation Team has been publishing high quality study materials since 2005, with a catalogue of over 145 titles, in English, French, Spanish and Chinese, as well as ESL curriculum for all levels.

To keep up with the industry changes, we update everything all the time!

And the best part?
With every purchase, you're helping people all over the world improve themselves and their education. So thank you in advance for supporting this mission with us! Together, we are truly making a difference in the lives of those often forgotten by the system.

Charities that we support

https://www.test-preparation.ca/charities-and-non-profits/

You have definitely come to the right place.
If you want to spend your valuable study time where it will help you the most - we've got you covered today and tomorrow.

Feedback

We welcome your feedback. Email us at feedback@test-preparation.ca with your comments and suggestions. We carefully review all suggestions and often incorporate reader suggestions into upcoming versions. As a Print on Demand Publisher, we update our products frequently.

Contents

8 **Getting Started**
   The Study Plan   9
   Making a Study Schedule   12

15 **Estimation**
   Practice Questions   18
   Answer Key   21

23 **Word Problems**
   Types of Word Problems   26
   Word Problem Practice
   `with Video Solutions   37
   Word Problem Practice   39
   Answer Key   47

56 **Fractions, Decimal and Percent**
   Fraction Tips, Tricks & Short-cuts   57
   Basic math Video Tutorials   57
   Fractions Decimal & Percent   57
   Decimal Tips, Tricks & Short-cuts   64
   Converting Decimals to Fractions   64
   Percent Tips, Tricks & Short-cuts   65
   Common Basic Math Mistakes   68
   Practice Questions   70
   Answer Key   77

82 **Arithmetic**
   Number Manipulation Strategy   82
   Practice Questions   86
   Answer Key   106

116 **Basic Algebra**
　　　　Solving One-Variable Linear Equations 116
　　　　Solving Two-Variable Linear Equations   17
　　　　Basic Algebra Solutions - Video Tutorials 119

124 **Geometry**
　　　　Basic Geometry   124
　　　　Pythagorean Geometry   132
　　　　Quadrilaterals   134
　　　　Practice Questions 140
　　　　Geometry Practice Questions   141

169 **Basic Math Multiple Choice**

175 **How to Study for a Math Test**

179 **How to Prepare for a Test**
　　　　The Strategy of Studying   181

184 **How to Take a Test**
　　　　Reading the Instructions   184
　　　　How to Take a Test - The Basics   186
　　　　In the Test Room 190
　　　　Avoid Anxiety Before a Test   196
　　　　Common Test-Taking Mistakes   198

201 **Conclusion**

202 **Online Resources**

Getting Started

CONGRATULATIONS! By deciding to take the basic math test, you have taken the first step toward a great future! Of course, there is no point in taking this important examination unless you intend to do your best to earn the highest grade that you possibly can. That means getting yourself organized and discovering the best approaches, methods and strategies to master the material. Yes, that will require real effort and dedication on your part, but if you are willing to focus your energy and devote the study time necessary, before you know it you will be opening that letter of acceptance to the school of your dreams.

We know that taking on a new endeavor can be scary, and it is easy to feel unsure of where to begin. That's where we come in. This study guide is designed to help you improve your test-taking skills, show you a few tricks of the trade and increase both your competency and confidence.

Basic Math Content

Basic Math

Basic operations with fractions, decimals and percent
Arithmetic
Estimation
Solve word problems

Basic Algebra

Solve equations with 1 variable

Basic Geometry

Lines and angles
Pythagorean geometry
Area, perimeter and volume

Non-Geometric Comparison

Number Manipulation

The Study Plan

Now that you have made the decision to take a math test, it's time to get started. Before you do another thing, you will need to figure out a plan of attack. The very best study tip is to start early! The longer the time period you devote to regular study practice, the more likely you will retain the material and be able to reach it quickly. If you thought that 1 x 20 is the same as 2 x 10, guess what? It really is not, when it comes to study time. Reviewing material for just an hour per day over the course of 20 days is far better than studying for two hours a day for only 10 days. The more often you revisit a particular piece of information, the better you will know it. Not only will your grasp and understanding improve, but your ability to reach into your brain and quickly and efficiently pull out the tidbit you need, will be greatly enhanced as well.

The great Chinese scholar and philosopher Confucius believed that true knowledge could be defined as knowing what you know and what you do not know. The first step in preparing for your test is to assess your strengths and weaknesses. You may already have an idea of what you know and what you do not know, but evaluating yourself for each of the math areas will clarify the details.

Making a Study Schedule

To make your study time the most productive, you will need to develop a study plan. The purpose of the plan is to organize all the bits of pieces of information in such a way that you will not feel overwhelmed. Rome was not built in a day, and learning everything you will need to know to pass your test is going to take time, too. Arranging the material you need to learn into manageable chunks is the best way to go. Each study session should make you feel as though you have accomplished your goal, and your goal is simply to learn what you planned to learn during that particular session. Try to organize the content in such a way that each study session builds on previous ones. That way, you will retain the information, be better able to reach it, and review the previous bits and pieces at the same time.

The Best Study Tip! The very best study tip is to start early! The longer you study regularly, the more you will retain and 'learn' the material. Studying for 1 hour per day for 20 days is far better than studying for 2 hours for 10 days.

What don't you know?

The first step is to assess your strengths and weaknesses. You may already have an idea of where your weaknesses are, or you can take our Self-assessment modules for each of the areas, math, English, science and reading.

Below is a table to assess your exam readiness in each content area. You can fill this in now, and correct if necessary after completing the self-assessments, or fill it in after you have taken the self-assessments.

Getting Started

Exam Readiness Assessment

Computation and Problem Solving	Rate 1 - 5
Basic operations with fractions, decimals and percent	
Arithmetic	
Basic operations with positive and negative numbers	
Estimation	
Solve word problems	
Basic Geometry	
Area, Perimeter and Volume	
Lines and Angles	
Pythagorean Geometry	
Basic Algebra	
Solve 1 variable equations	
Non-Geometric Comparison	
Number Manipulation	

Making a Study Schedule

The key to a study plan is to divide the material you need to learn into manageable size and learn it, while at the same time reviewing the material that you already know.

Using the table above, any scores of 3 or below, you need to spend time learning, going over, and practicing this subject area. A score of 4 means you need to review the material, but you don't have to spend time re-learning. A score of 5 and you are OK with just an occasional review before the exam.

A score of 0 or 1 means you really need to work on this area and should allocate the most time and the highest priority. Some students prefer a 5-day plan and others a 10-day plan. It also depends on how much time until the exam.

Here is an example of a 5-day plan based on an example from the table above:

> **Fractions, Decimals, Percent:** 1 Study 1 hour everyday – review on last day
> **Estimation:** 3 Study 1 hour for 2 days then ½ hour a day, then review
> **Number Manipulation:** 4 Review every second day
> **Area, Perimeter & Volume:** 2 Study 1 hour on the first day – then ½ hour everyday
> **Algebra:** 5 Review for ½ hour every other day
> **Word Problems:** 5 very confident – review a few times.

Using this example, here is a sample study plan which you can adapt to your own situation:

Getting Started

Day	Subject	Time
Monday		
Study	Fractions, Decimals, Percent	1 hour
Study	Area, Perimeter & Volume	1 hour
	½ hour break	
Study	Estimation	1 hour
Review	Word Problems	½ hour
Tuesday		
Study	Fractions, Decimals, Percent	1 hour
Study	Area, Perimeter & Volume	½ hour
	½ hour break	
Study	Data Interpretation	½ hour
Review	Number Manipulation	½ hour
Review	Grammar	½ hour
Wednesday		
Study	Fractions, Decimals, Percent	1 hour
Study	Area, Perimeter & Volume	½ hour
	½ hour break	
Study	Estimation	½ hour
Review	Word Problems	½ hour
Thursday		
Study	Fractions, Decimals, Percent	½ hour
Study	Area, Perimeter & Volume	½ hour
Review	Estimation	½ hour

	½ hour break	
Review	Estimation	½ hour
Review	Number Manipulation	½ hour
Friday		
Review	Fractions, Decimals, Percent	½ hour
Review	Area, Perimeter & Volume	½ hour
Review	Estimation	½ hour
	½ hour break	
Review	Number Manipulation	½ hour
Review	Area, Perimeter & Volume	½ hour

Estimation Strategy

Strategy 1: Break it down

Estimate 105 X 8

    a. 840
    b. 922
    c. 880
    d. 860

Answer: A
Break 105 into 2 parts: 100 and 5, then multiple both by 8, and add.

100 X 8 is easy - 100 X 8 = 800. And, 5 X 8 = 40, adding gives the answer, 840, Choice A

Strategy 2: Use base 10

Estimate 1050 x 128

    a. 210,000
    b. 200,000
    c. 21,000
    d. 130,000

1050 and 128 are difficult to multiply in your head so take 1000 and 100 - add two zeros to 1000 for the answer, 100,000. Because we rounded down to 1000 and 100, this estimate will be lower than the actual answer.

Looking at the choices, A, B and C can all be eliminated as too small (C) or too large (A and B) so the answer much be D.

Confirming with a calculator, 1050 X 128 = 134,400.

Answer Sheet

	A	B	C	D
1	○	○	○	○
2	○	○	○	○
3	○	○	○	○
4	○	○	○	○
5	○	○	○	○
6	○	○	○	○
7	○	○	○	○
8	○	○	○	○
9	○	○	○	○
10	○	○	○	○

Practice Questions

1. Brad has agreed to buy everyone a Coke. Each drink costs $1.89, and there are 5 friends. Estimate Brad's cost.

    a. $7
    b. $8
    c. $10
    d. $12

2. What is the best approximate solution for 1.135 - 113.5?

    a. -110
    b. 100
    c. -90
    d. 110

3. Estimate 16 x 230

    a. 31,000
    b. 301,000
    c. 3,100
    d. 3,000,000

4. Estimate 215 x 65

    a. 1,350
    b. 13,500
    c. 103,500
    d. 3,500

5. Estimate 2009 x 108

   a. 110,000
   b. 200,000
   c. 21,000
   d. 210,000

6. Estimate 46,227 + 101,032

   a. 14,700
   b. 147,000
   c. 14,700,000
   d. 104,700

7. Estimate 4,210,987 – 210,078

   a. 4,000,000
   b. 40,000,000
   c. 400,000
   d. 40,000

8. Estimate 5205 / 25

   a. 108
   b. 308
   c. 208
   d. 408

9. Estimate 2045 / 15

   a. 140
   b. 1500
   c. 105
   d. 350

10. Estimate 136 / 12

    a. 10
    b. 11
    c. 12
    d. 13

— Estimation —

Answer Key

1. C
If there are 5 friends, and each drink costs $1.89, we can round up to $2 per drink and estimate the total cost at, 5 X $2 = $10. The actual cost is 5 X $1.89 = $9.45.

2. A
1.135 -113.5 = -112.37. Best approximate = -110

3. C
16 X 230 = 3680
To estimate, break 16 into 10 and 6. 10 * 230 = 2300, and 6 * 230 will be about half that 1150 (actually 1380)

For an approximation, 2300 + 1150 = 3450. The only choice is choice C, 3100.

4. B
215 X 65 = 13975
Choices A (1,350) and D (3,500) can be eliminated right away as they are too small. Choice C (103,500) is too large and can be eliminated, leaving only choice B.

5. D
2009 * 108 = 216972
To estimate, use 2000 * 100, or add 2 zeros to 2000 for 200,000. The only choice that is close is choice D (210,000)

6. B
46,227 + 101,032 = 147,032
To estimate, use 50,000 + 100,000 = 150,000. The only choice close to 150,000 is choice B.

7. A
4,210,987 − 210,078
To estimate, use 4,000,000 and 200,000 = 3,800,000 and find the choice that is closest - choice A, 4,000,000

8. C
5205/25 = 208.2
To estimate, start with easy number, 1000 and divide by 25 = 40. Or, take 100 and divide by 25 = 4 and multiple by 10 get 1000/25.

so, if 1000/25 = 40, then multiple by 5 for 5000/25 = 200. The only choice close to 200 is choice C, 208.

9. A
2045 / 15 = 136.3333
To estimate, use 1000 / 10 = 100, which eliminates choice B and D. The two remaining choices are A and 140 and C 105. Looking at choice C, estimate 100 * 15 = 1500, which is too low, so the answer is choice A.

10. B
136/12 = 11.3333
To estimate, remember 12 X 12 = 144, so choices C and D can be eliminated. Choice A can be eliminated since 10 * 12 = 120, so the answer must be choice B

How to Solve Word Problems

How to Solve Word Problems

Do you know what the biggest tip for solving word problems is?

Practice regularly and systematically.

Sounds simple and easy right? Yes it is, and yes it really does work.

Word problems are a way of thinking and require you to translate a real-world problem into mathematical terms.

Some math teachers say that learning how to think mathematically is the main reason for teaching word problems.

So what does that mean?

Studying word problems and math in general requires a logical and mathematical frame of mind. The only way you can get this is by practicing regularly, which means every day.

It is critical that you practice word problems every day for the 5 days before the exam as the absolute minimum.

If you practice and miss a day, you have lost

the mathematical frame of mind and the benefit of your previous practice is gone. You must start all over again.

Everything is important.

All the information given in the problem has some purpose. There is no unnecessary information! Word problems are typically around 50 words in 2 or 3 sentences.

Often, the relationships are complicated. To explain everything, every word counts.

Make sure that you use every piece of information.

7 steps to solving word problems

Step 1 – Read through the problem at least three times. The first reading should be a quick scan, and the next two readings should be done slowly to find answers to these questions:

> What does the problem ask? (Usually located at the end)

Mark all information and underline all important words or phrases.

Step 2 – Draw a picture. Use arrows, circles, lines, whatever works for you.   This makes the problem real.

A favorite word problem is something like, 1 train leaves Station A travelling at 100 km/hr and another train leaves Station B travelling at 60 km/hr. …

Draw a line, the two stations, and the two trains at either end.

Depending on the question, make a table with a blank portion to show information you don't know.

Step 3 – Assign a single letter to represent each unknown.

You may want to note the unknown that each letter represents so you don't get confused.

Step 4 – Translate the information into an equation.

Remember that the main problem with word problems is that they are not expressed in regular math equations. Your ability to identify correctly the variables and translate the information into an equation determines your ability to solve the problem.

Step 5 – Check the equation to see if it looks like regular equations that you are used to seeing and whether it looks sensible.

Does the equation appear to represent the information in the question? Take note that you may need to rewrite some formulas needed to solve the word problem equation.

Step 6 – Use algebra rules to solve the equation.

Simplify each side of the equation by removing parentheses and combining like terms.

Use addition or subtraction to isolate the variable term on one side of the equation. If a number crosses to the other side of the equation, the sign changes to the opposite -- for example positive to negative.

Use multiplication or division to solve for the variable. What you to once side of the equation you must do for the other.

Where there are multiple unknowns you will need to use elimination or substitution methods to resolve all the equations.

Step 7 – Check your final answers to see if they make sense with the information given in the problem.

For example, if the word problem involves a discount,

the final price should be less or if a product was taxed then the final answer has to cost more.

Types of Word Problems

Word problems can be classified into 12 types. Below are examples of each type with a complete solution. Some types of word problems can be solved quickly using multiple choice strategies and some cannot. Always look for ways to estimate the answer and then eliminate choices.

1. Age

A girl is 10 years older than her brother. By next year, she will be twice the age of her brother. What are their ages now?

    a. 25, 15
    b. 19, 9
    c. 21, 11
    d. 29, 19

Solution: B

We will assume that the girl's age is "a" and her brother's age is "b." This means that based on the information in the first sentence,
$a = 10 + b$

Next year, she will be twice her brother's age, which gives, $a + 1 = 2(b + 1)$

We need to solve for one unknown factor and then use the answer to solve for the other. To do this we substitute the value of "a" from the first equation into the second equation. This gives

$10 + b + 1 = 2b + 2$

11 + b = 2b + 2
11 − 2 = 2b − b
b = 9

9 = b this means that her brother is 9 years old. Solving for the girl's age in the first equation gives a = 10 + 9. a = 19 the girl is aged 19. So, the girl is aged 19 and the boy is 9

2. Distance or Speed

Two boats travel down a river towards the same destination, starting at the same time. One is traveling at 52 km/hr, and the other boat at 43 km/hr. How far apart will they be after 40 minutes?

    a. 46.67 km

    b. 19.23 km

    c. 6.4 km

    d. 14.39 km

Solution: C

After 40 minutes, the first boat will have traveled = 52 km/hr x 40 minutes/60 minutes = 34.7 km
After 40 minutes, the second boat will have traveled = 43 km/hr x 40/60 minutes = 28.66 km
Difference between the two boats will be 34.7 km − 28.66 km = 6.04 km.

Multiple Choice Strategy

First estimate the answer. The first boat is traveling 9 km. faster than the second, for 40 minutes, which is 2/3 of an hour. 2/3 of 9 = 6, as a rough guess of the distance apart.

Choices A, B and D can be eliminated right away.

3. Ratio

The instructions in a cookbook state that 700 grams of flour must be mixed in 100 ml of water, and 0.90 grams of salt added. A cook however has just 325 grams of flour. What is the quantity of water and salt that he should use?

    a. 0.41 grams and 46.4 ml
    b. 0.45 grams and 49.3 ml
    c. 0.39 grams and 39.8 ml
    d. 0.25 grams and 40.1 ml

Solution: A

The Cookbook states 700 grams of flour, but the cook only has 325. The first step is to determine the percentage of flour he has 325/700 x 100 = 46.4%
That means that 46.4% of all other items must also be used.
46.4% of 100 = 46.4 ml of water
46.4% of 0.90 = 0.41 grams of salt.

Multiple Choice Strategy

The recipe calls for 700 grams of flour but the cook only has 325, which is just less than half, the quantity of water and salt are going to be about half.

Choices C and D can be eliminated right away. Choice B is very close so be careful. Looking closely at choice B, it is exactly half, and since 325 is slightly less than half of 700, it can't be correct.

Choice A is correct.

4. Percent

An agent received $6,685 as his commission for selling a property. If his commission was 13% of the selling price, how much was the property?

    a. $68,825
    b. $121,850
    c. $49,025
    d. $51,423

Solution: D

Let's assume that the property price is x. That means from the information given, 13% of x = 6,685
Solve for x,

x = 6685 x 100/13 = $51,423

Multiple Choice Strategy

The commission, 13%, is just over 10%, which is easier to work with. Round up $6685 to $6700, and multiple by 10 for an approximate answer. 10 X 6700 = $67,000. You can do this in your head. Choice B is much too big and can be eliminated. Choice C is too small and can be eliminated. Choices A and D are left and good possibilities.

Do the calculations to make the final choice.

5. Sales & Profit

A store owner buys merchandise for $21,045. He transports them for $3,905 and pays his staff $1,450 to stock the merchandise on his shelves. If he does not incur further costs, how much does he need to sell the items to make $5,000 profit?

    a. $32,500
    b. $29,350
    c. $32,400
    d. $31,400

Solution: D

Total cost of the items is $21,045 + $3,905 + $1,450 = $26,400

Total cost is now $26,400 + $5000 profit = $31,400

Multiple Choice Strategy

Round off and add the numbers up in your head quickly. 21,000 + 4,000 + 1500 = 26500. Add in 5000 profit for a total of 31500.

Choice B is too small and can be eliminated. Choice C and Choice A are too large and can be eliminated.

6. Tax/Income

A woman earns $42,000 per month and pays 5% tax on her monthly income. If the Government increases her monthly taxes by $1,500, what is her income after tax?

 a. $38,400
 b. $36,050
 c. $40,500
 d. $39, 500

Solution: A

Initial tax on income was 5/100 x 42,000 = $2,100
$1,500 was added to the tax to give $2,100 + 1,500 = $3,600
Income after tax is $42,000 - $3,600 = $38,400

7. Simple Interest Word Problems

Simple interest is one type of interest problems. There are always four variables of any simple interest equation. With simple interest, you would be given three of these variables and be asked to solve for one unknown variable. With more complex interest problems, you would have to solve for multiple variables.

The four variables of simple interest are:

> P – Principal which refers to the original amount of money put in the account
> I – Interest or the amount of money earned as interest
> r – Rate or interest rate. This MUST ALWAYS be in decimal format and not in percentage
> t – Time or the amount of time the money is kept in the account to earn interest

The formula for simple interest is I = P x r x t

Example 1

A customer deposits $1,000 in a savings account with a bank that offers 2% interest. How much interest will be earned after 4 years?

For this problem, there are 3 variables as expected.

P = $1,000
t = 4 years
r = 2%
I = ?

Before we can begin solving for I using the simple interest formula, we need to first convert the rate from percentage to decimal.

2% = 2/100 = 0.02
Now we can use the formula: I = P x r x t

I = 1,000 x 0.02 x 4 = 80
This means that the $1,000 would have earned an interest of $80 after 4 years. The total in the account after 4 years will thus be principal + interest earned, or 1,000 + 80 = $1,080

Example 2

Sandra deposits $1400 in a savings account with a bank at 5% interest. How long will she have to leave the money in the bank to earn $420 as interest to buy a second-hand car?

In this example, the given information is:
I = $420
P = $1,400
r - 5%
t - ?

As usual, first we convert the rate from percentage to decimal

5% = 5/100 = 0.05

Next, we plug in the variables we know into the simple interest formula - I = P x r x t

420 = 1,400 x 0.05 x t
420 = 70 x t
420 = 70t
t = 420/70
t = 6

Sandra will have to leave her $1,400 in the bank for 6 years to earn her an interest of $420 at a rate of 5%.

Other important simple interest formula to remember are below. To use these formula, do not convert r (rate) to decimal.

P = 100 x interest/ r x t
r = 100 x interest/p x t
t = 100 x interest/ p x r

8. Averaging

The average weight of 10 books is 54 grams. 2 more books were added and the average weight became 55.4. If one of the 2 new books added weighed 62.8 g, what is the weight of the other?

    a. 44.7 g
    b. 67.4 g
    c. 62 g
    d. 52 g

Solution: C

Total weight of 10 books with average 54 grams will be
= 1 0 × 54 = 540 g

Total weight of 12 books with average 55.4 will be
= 55.4 × 12 = 664.8 g

Total weight of the remaining 2 will be
= 664.8 − 540 = 124.8 g

If one weighs 62.8, the weight of the other will be
= 124.8 g − 62.8 g = 62 g

Multiple Choice Strategy

Averaging problems can be estimated by looking at which direction the average goes. If additional items are added and the average goes up, the new items much be greater than the average. If the average goes down after new items are added, the new items must be less than the average.

Here, the average is 54 grams and 2 books are added which increases the average to 55.4, so the new books must weight more than 54 grams.
Choices A and D can be eliminated right away.

9. Probability

A bag contains 15 marbles of various colors. If 3 marbles are white, 5 are red and the rest are black, what is the probability of randomly picking out a black marble from the bag?

    a. 7/15
    b. 3/15
    c. 1/5
    d. 4/15

Solution: A

Total marbles = 15
Number of black marbles = 15 − (3 + 5) = 7
Probability of picking out a black marble = 7/15

10. Two Variables

A company paid a total of $2850 to book for 6 single rooms and 4 double rooms in an hotel for one night. Another company paid $3185 to book for 13 single rooms for one night in the same hotel. What is the cost for single and double rooms in that hotel?

    a. single= $250 and double = $345
    b. single= $254 and double = $350
    c. single = $245 and double = $305
    d. single = $245 and double = $345

Solution: D

We can determine the price of single rooms from the information given of the second company. 13 single rooms = 3185.
One single room = 3185 / 13 = 245

The first company paid for 6 single rooms at $245. 245 x 6 = $1470

Total amount paid for 4 double rooms by first company = $2850 - $1470 = $1380

Cost per double room = 1380 / 4 = $345

11. Geometry

The length of a rectangle is 5 in. more than its width. The perimeter of the rectangle is 26 in. What is the width and length of the rectangle?

    a. width = 6 inches, Length = 9 inches

    b. width = 4 inches, Length = 9 inches

    c. width =4 inches, Length = 5 inches

    d. width = 6 inches, Length = 11 inches

Solution: B

Formula for perimeter of a rectangle is 2(L + W)
p = 26, so 2(L + W) = p

The length is 5 inches more than the width, so
2(w + 5) + 2w = 26
2w + 10 + 2w = 26
2w + 2w = 26 - 10
4w = 16

W = 16/4 = 4 inches

L is 5 inches more than w, so L = 5 + 4 = 9 inches.

12. Totals and fractions

A basket contains 125 oranges, mangoes and apples. If 3/5 of the fruits in the basket are mangoes and only 2/5 of the mangoes are ripe, how many ripe mangoes are there in the basket?

    a. 30
    b. 68
    c. 55
    d. 47

Solution: A
Number of mangoes in the basket is 3/5 x 125 = 75
Number of ripe mangoes = 2/5 x 75 = 30

Word Problem Practice with Video Solutions

HTTPS://YOUTU.BE/6XWA6FO6YCE

Adult Math Workbook

Answer Sheet

	A	B	C	D	E		A	B	C	D	E
1	○	○	○	○	○	21	○	○	○	○	○
2	○	○	○	○	○	22	○	○	○	○	○
3	○	○	○	○	○	23	○	○	○	○	○
4	○	○	○	○	○	24	○	○	○	○	○
5	○	○	○	○	○	25	○	○	○	○	○
6	○	○	○	○	○
7	○	○	○	○	○
8	○	○	○	○	○
9	○	○	○	○	○
10	○	○	○	○	○
11	○	○	○	○	○
12	○	○	○	○	○
13	○	○	○	○	○
14	○	○	○	○	○
15	○	○	○	○	○
16	○	○	○	○	○
17	○	○	○	○	○
18	○	○	○	○	○
19	○	○	○	○	○
20	○	○	○	○	○

Word Problem Practice

1. Translate the following into an equation: Five greater than 3 times a number.

    a. 3X + 5
    b. 5X + 3
    c. (5 + 3)X
    d. 5(3 + X)

2. Translate the following into an equation: three plus a number times 7 equals 42.

    a. 7(3 + X) = 42
    b. 3(X + 7) = 42
    c. 3X + 7 = 42
    d. (3 + 7)X = 42

3. Translate the following into an equation: 2 + a number divided by 7.

    a. (2 + X)/7
    b. (7 + X)/2
    c. (2 + 7)/X
    d. 2/(7 + X)

4. Translate the following into an equation: six times a number plus five.

    a. 6X + 5
    b. 6(X+5)
    c. 5X + 6
    d. (6 * 5) + 5

5. A box contains 7 black pencils and 28 blue ones. What is the ratio between the black and blue pens?

    a. 1:4
    b. 2:7
    c. 1:8
    d. 1:9

6. The manager of a weaving factory estimates that if 10 machines run at 100% efficiency for 8 hours, they will produce 1450 meters of cloth. Due to some technical problems, 4 machines run of 95% efficiency and the remaining 6 at 90% efficiency. How many meters of cloth can these machines will produce in 8 hours?

    a. 1334 meters
    b. 1310 meters
    c. 1300 meters
    d. 1285 meters

7. In a local election at polling station A, 945 voters cast their vote out of 1270 registered voters. At polling station B, 860 cast their vote out of 1050 registered voters and at station C, 1210 cast their vote out of 1440 registered voters. What is the total turnout from all three polling stations?

    a. 70%
    b. 74%
    c. 76%
    d. 80%

Word Problems

8. If Lynn can type a page in p minutes, what portion of the page can she do in 5 minutes?

    a. p/5
    b. p - 5
    c. p + 5
    d. 5/p

9. If Sally can paint a house in 4 hours, and John can paint the same house in 6 hours, how long will it take for both to paint a house?

    a. 2 hours and 24 minutes
    b. 3 hours and 12 minutes
    c. 3 hours and 44 minutes
    d. 4 hours and 10 minutes

10. Employees of a discount appliance store receive an additional 20% off the lowest price on any item. If an employee purchases a dishwasher during a 15% off sale, how much will he pay if the dishwasher originally cost $450?

    a. $280.90
    b. $287.00
    c. $292.50
    d. $306.00

11. The sale price of a car is $12,590, which is 20% off the original price. What is the original price?

    a. $14,310.40
    b. $14,990.90
    c. $15,108.00
    d. $15,737.50

12. Richard gives 's' amount of salary to each of his 'n' employees weekly. If he has 'x' amount of money, how many days he can employ these 'n' employees.

    a. sx/7n
    b. 7x/nx
    c. nx/7s
    d. 7x/ns

13. A distributor purchased 550 pounds of potatoes for $165. He distributed these at a rate of $6.4 per 20 pounds to 15 shops, $3.4 per 10 pounds to 12 shops and the remainder at $1.8 per 5 pounds. If his total distribution cost is $10, what will his profit be?

    a. $10.40
    b. $8.60
    c. $14.90
    d. $23.40

14. How much pay does Mr. Johnson receive if he gives half of his pay to his family, $250 to his landlord, and has exactly 3/7 of his pay left over?

    a. $3600
    b. $3500
    c. $2800
    d. $1750

Word Problems

15. The cost of waterproofing canvas is .50 a square yard. What's the total cost for waterproofing a canvas truck cover that is 15' x 24'?

    a. $18.00
    b. $6.67
    c. $180.00
    d. $20.00

16. The price of a book went from $20 to $25. What percent did the price increase?

    a. 5%
    b. 10%
    c. 20%
    d. 25%

17. In the time required to serve 43 customers, a server breaks 2 glasses and slips 5 times. The next day, the same server breaks 10 glasses. Assuming that glasses broken is proportional to customers served, how many customers did she serve?

    a. 25
    b. 43
    c. 86
    d. 215

18. A square lawn has an area of 62,500 square meters. What is the cost of building fence around it at a rate of $5.5 per meter?

    a. $4000
    b. $4500
    c. $5000
    d. $5500

19. Susan wants to buy a leather jacket that costs $545.00 and is on sale for 10% off. What is the approximate cost?

    a. $525
    b. $450
    c. $475
    d. $500

20. Sarah weighs 25 pounds more than Tony. If together they weigh 205 pounds, how much does Sarah weigh in kilograms? Assume 1 pound = 0.4535 kilograms.

    a. 41
    b. 48
    c. 50
    d. 52

— Word Problems —

21. A man buys an item for $420 and has a balance of $3000.00. How much did he have before his purchase?

    a. $2,580
    b. $3,420
    c. $2,420
    d. $342

22. The average weight of 13 students in a class of 15 (two were absent that day) is 42 kg. When the remaining two are weighed, the average became 42.7 kg. If one of the remaining students weighs 48 kg., how much does the other weigh?

    a. 44.7 kg.
    b. 45.6 kg.
    c. 46.5 kg.
    d. 47.4 kg.

23. The total expense of building a fence around a square-shaped field is $2000 at a rate of $5 per meter. What is the length of one side?

    a. 40 meters
    b. 80 meters
    c. 100 meters
    d. 320 meters

24. There were some oranges in a basket. By adding 8/5 of the total to the basket, the new total is 130. How many oranges were in the basket?

    a. 60

    b. 50

    c. 40

    d. 35

25. A person earns $25,000 per month and pays $9,000 income tax per year. The Government increased income tax by 0.5% per month and his monthly earning was increased $11,000. How much more income tax will he pay per month?

    a. $1260

    b. $1050

    c. $750

    d. $510

Answer Key

Part 1 - Equation Translation

1. A
Five greater than 3 times a number.
5 + 3 times a number.
3X + 5

2. A
Three plus a number times 7 equals 42.
Let X be the number.
(3 + X) times 7 = 42
7(3 + X) = 42

3. A
2 + a number divided by 7.
(2 + X) divided by 7.
(2 + X)/7

4. B
Six times a number plus five is the same as saying six times (a number plus five). Or,
6 * (a number plus five). Let X be the number so,
6(X + 5).

5. A
The ratio between black and blue pens is 7 to 28 or 7:28. Bring to the lowest terms by dividing both sides by 7 gives 1:4.

6. A
At 100% efficiency 1 machine produces 1450/10 = 145 m of cloth.

At 95% efficiency, 4 machines produce 4 * 145 * 95/100 = 551 m of cloth.

At 90% efficiency, 6 machines produce 6 * 145 * 90/100 =

783 m of cloth.

Total cloth produced by all 10 machines = 551 + 783 = 1334 m

Since the information provided and the question are based on 8 hours, we did not need to use time to reach the answer.

7. D
https://youtu.be/Es3Yg5pfYeY

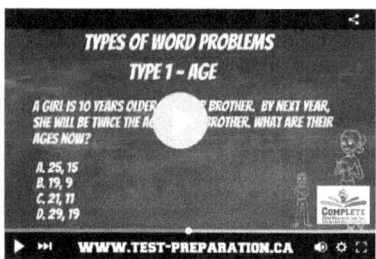

To find the total turnout in all three polling stations, we need to proportion the number of voters to the number of all registered voters.

Number of total voters = 945 + 860 + 1210 = 3015

Number of total registered voters = 1270 + 1050 + 1440 = 3760

Percentage turnout over all three polling stations = 3015 * 100/3760 = 80.19%

Checking the answers, we round 80.19 to the nearest whole number: 80%

8. D
https://youtu.be/syDAMxmkYgY

This is a simple direct proportion problem:
If Lynn can type 1 page in p minutes,

she can type x pages in 5 minutes

We do cross multiplication: x * p = 5 * 1

Then,

x = 5/p

9. A
This is an inverse ratio problem.

1/x = 1/a + 1/b where a is the time Sally can paint a house, b is the time John can paint a house, x is the time Sally and John can together paint a house.

So,

1/x = 1/4 + 1/6 ... We use the least common multiple in the denominator that is 24:

1/x = 6/24 + 4/24

1/x = 10/24

x = 24/10

x = 2.4 hours.

In other words; 2 hours + 0.4 hours = 2 hours + 0.4•60 minutes

= 2 hours 24 minutes

10. D
https://youtu.be/l_FaJPJzepE

The cost of the dishwasher = $450

15% discount amount = 450•15/100 = $67.5

The discounted price = 450 − 67.5 = $382.5

20% additional discount amount on lowest price = 382.5•20/100 = $76.5

So, the final discounted price = 382.5 - 76.5 = $306.00

11. D
Original price = x,
80/100 = 12590/X,
80X = 1259000,
X = 15,737.50.

12. D
https://youtu.be/EF92e6V4mAA

We are given that each of the n employees earns s amount of salary weekly. This means that one employee earns s salary weekly. So; Richard has 'ns' amount of money to employ n

employees for a week.

We are asked to find the number of days n employees can be employed with x amount of money. We can do simple direct proportion:

If Richard can employ n employees for 7 days with 'ns' amount of money,

Richard can employ n employees for y days with x amount of money ... y is the number of days we need to find.

We can cross multiply:

y = (x * 7)/(ns)

y = 7x/ns

13. B
The distribution is done at three different rates and in three different amounts:

$6.4 per 20 kilograms to 15 shops ... 20•15 = 300 kilograms distributed

$3.4 per 10 kilograms to 12 shops ... 10•12 = 120 kilograms distributed

550 - (300 + 120) = 550 - 420 = 130 kilograms left. This amount is distributed in 5 kilogram portions. So, this means that there are 130/5 = 26 shops.

$1.8 per 130 kilograms.

We need to find the amount he earned overall these distributions.

$6.4 per 20 kilograms : 6.4 * 15 = $96 for 300 kilograms

$3.4 per 10 kilograms : 3.4 *12 = $40.8 for 120 kilograms

$1.8 per 5 kilograms : 1.8 * 26 = $46.8 for 130 kilograms

So, he earned 96 + 40.8 + 46.8 = $ 183.6

The total distribution cost is given as $10

The profit is found by: Money earned - money spent ... It is important to remember that he bought 550 kilograms of potatoes for $165 at the beginning:

Profit = 183.6 - 10 - 165 = $8.6

14. B

We check the fractions taking place in the question. We see that there is a "half" (that is 1/2) and 3/7. So, we multiply the denominators of these fractions to decide how to name the total money. We say that Mr. Johnson has 14x at the beginning; he gives half of this, meaning 7x, to his family. $250 to his landlord. He has 3/7 of his money left. 3/7 of 14x is equal to:

14x * (3/7) = 6x

So,

Spent money is: 7x + 250

Unspent money is: 6x

Total money is: 14x

Write an equation: total money = spent money + unspent money

14x = 7x + 250 + 6x

14x - 7x - 6x = 250

x = 250

We are asked to find the total money that is 14x:

Word Problems

14x = 14 * 250 = $3500

15. D
First calculate total square feet, which is 15•24 = 360 ft2. Next, convert this value to square yards, (1 yards2 = 9 ft2) which is 360/9 = 40 yards2. At $0.50 per square yard, the total cost is 40 * 0.50 = $20.

16. D
Price increased by $5 ($25-$20). To calculate the percent increase:
5/20 = X/100
500 = 20X
X = 500/20
X = 25%

17. D
2 glasses are broken for 43 customers so 1 glass breaks for every 43/2 customers served, therefore 10 glasses implies (43/2) * 10 = 215 customers.

18. D
As the lawn is square, the length of one side will be the square root of the area. √62,500 = 250 meters. So, the perimeter is found by 4 times the length of the side of the square:

250 * 4 = 1000 meters.

Since each meter costs $5.5, the total cost of the fence will be 1000 * 5.5 = $5,500.

19. D
The question asks for approximate cost, so work with round numbers. The jacket costs $545.00 so we can round up to $550. 10% of $550 is 55. We can round down to $50, which is easier to work with. $550 - $50 is $500. The jacket will cost about $500.

The actual cost will be 10% X 545 = $54.50
545 – 54.50 = $490.50

20. D

Let us denote Sarah's weight by "x." Then, since she weighs 25 pounds more than Tony, so he will be x - 25. They together weigh 205 pounds which means that the sum of the two representations will be equal to 205:

Sarah : x

Tony : x - 25

x + (x - 25) = 205 ... by arranging this equation we have:

x + x - 25 = 205

2x - 25 = 205 ... we add 25 to each side to have x term alone:

2x - 25 + 25 = 205 + 25

2x = 230

x = 230/2

x = 115 pounds → Sarah weighs 115 pounds. Since 1 pound is 0.4535 kilograms, we need to multiply 115 by 0.4535 to have her weight in kilograms:

x = 115 * 0.4535 = 52.1525 kilograms → this is equal to 52 when rounded to the nearest whole number.

21. B
(Amount Spent) $420 + $3000 (Balance) = $3,420.00

― Word Problems ―

22. C

Total weight of 13 students with average 42 will be = 42 * 13 = 546 kg.

The total weight of the remaining 2 will be found by subtracting the total weight of 13 students from the total weight of 15 students: 640.5 - 546 = 94.5 kg.

94.5 = the total weight of two students. One of these students weigh 48 kg, so;

The weight of the other will be = 94.5 – 48 = 46.5 kg

23. C

Total expense is $2000 and we are informed that $5 is spent per meter. Combining these two information, we know that the total length of the fence is 2000/5 = 400 meters.

The fence is built around a square field. If one side of the square is "a," the perimeter of the square is "4a." Here, the perimeter is equal to 400 meters. So,

$400 = 4a$

$100 = a \rightarrow$ this means that one side of the square is equal to 100 meters

24. B

Let the number of oranges in the basket before additions = x
Then: $X + 8x/5 = 130$
$5x + 8x = 650$
$650 = 13x$
$X = 50$

25. D

The income tax per year is $9,000. So, the income tax per month is $9,000/12 = 750.

This person earns $25,000 per month and pays $750 income tax. We need to find the rate of the income tax:

Tax rate: $750 * 100/25,000 = 3\%$
Government increased this rate by 0.5% so it became 3.5%.

The income of the person per month is increased $11,000 so it became:

$25,000 + $11,000 = $36,000.

The new monthly income tax is: $36,000 * 3.5/100 = 1260.

Amount of increase in tax per month is:
$1260 - $750 = $510.

Fraction Tips, Tricks and Shortcuts

Basic math Video Tutorials

https://www.test-preparation.ca/basic-math-video-tutorials/
https://test-preparation.ca/basic-math-practice-test-questions/

Fraction Tips, Tricks and Short-cuts

When you are writing an exam, time is precious, so anything you can do to answer questions faster is a real advantage.

Here are some ideas, Short-cuts, tips and tricks that can speed up answering fraction problems.

Remember that a fraction is just a number which names a portion of something. For instance, instead of having a whole pie, a fraction says you have a part of a pie--such as a half of one or a fourth of one.

Two numbers make up a fraction. The number on top is the numerator. The number on the bottom is the denominator.

To remember which is which, just remember that "denominator" and "down" both start with a "d." And the "downstairs" number is the denominator. So for instance, in ½, the numerator is 1, and the denominator (or "downstairs") number is 2.

Adding Fractions

It's easy to add two fractions if they have the same denominator. Just add the digits on top and leave the bottom one the same: 1/10 + 6/10 = 7/10.

It's the same with subtracting fractions with the same denominator: 7/10 - 6/10 = 1/10.

Adding and subtracting fractions with different denominators is a little more complicated.

First, you have to arrange the fractions so they have the same denominators.

The easiest way to do this is to multiply the denominators: For 2/5 + 1/2 multiply 5 by 2. Now you have a denominator of 10.

But now you have to change the top numbers too. Since you multiplied the 5 in 2/5 by 2, you also multiply the 2 by 2, to get 4. So the first fraction is now 4/10.

In the second fraction, you multiplied the denominator by 5, you have to multiply the numerator by 5 also, to get 5/10.

Now you have 4/10 + 5/10 and you can add 5 and 4 to get 9/10.

Simplest Form

To reduce a fraction to its simplest form, you have to arrange the numerator and denominator so the only common factor is 1.

Fractions, Decimals and Percent

Think of it this way:

Let's take an example: The fraction 2/10.

This is not reduced to its simplest terms because there is a number that will divide evenly into both: 2. We want to make it so that the only number that will divide evenly into both is 1.

Divide the top and bottom by 2 to get the new, reduced fraction - 1/5.

Multiplying Fractions

This is the easiest of all: Just multiply the two top numbers and then multiply the two bottom numbers.

Here is an example,

2/5 X 2/3

First, multiply the numerators: 2 X 2 = 4

then multiply the denominators: 5 X 3 = 15

Your answer is 4/15.

Dividing Fractions

Dividing fractions is easy if you remember a simple trick - first turn the second fraction upside down - then multiply!

Here is an example:

7/8 X 1/2

Turn the second fraction upside down:

7/8 X 2/1

then multiply:
(7 X 2) / (8 X 1) = 14/8

Converting Fractions to Decimals

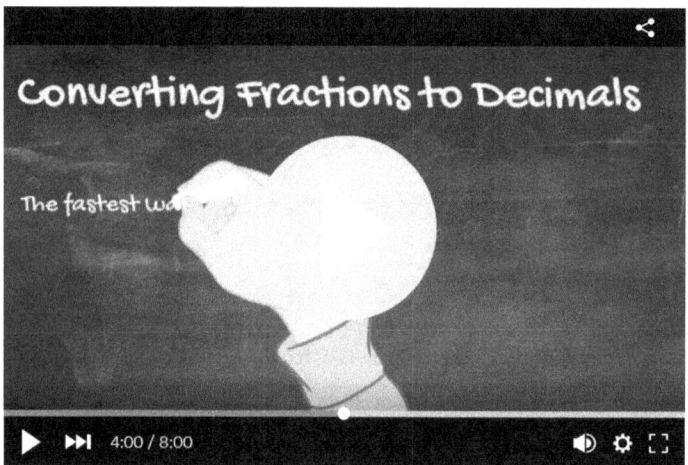

https://youtu.be/0OA9yXAc4fY

There are a couple of ways to convert fractions to decimals. The first, which is the fastest -- is to memorize some basic fraction facts.

1/100 is "one hundredth," expressed as a decimal, it's .01.

1/50 is "two hundredths," expressed as a decimal, it's .02.

1/25 is "one twenty-fifth" or "four hundredths," expressed as a decimal, it's .04.

1/20 is "one twentieth" or ""five hundredths," expressed as a decimal, it's .05.

— Fractions, Decimals and Percent —

1/10 is "one tenth," expressed as a decimal, it's .1.

1/8 is "one eighth," or "one hundred twenty-five thousandths," expressed as a decimal, it's .125.

1/5 is "one fifth," or "two tenths," expressed as a decimal, it's .2.

1/4 is "one fourth" or "twenty-five hundredths," expressed as a decimal, it's .25.

1/3 is "one third" or "thirty-three hundredths," expressed as a decimal, it's .33.

1/2 is "one half" or "five tenths," expressed as a decimal, it's .5.

3/4 is "three fourths," or "seventy-five hundredths," expressed as a decimal, it's .75.

Of course, if you're no good at memorization, another good technique for converting a fraction to a decimal is to manipulate it so that the fraction's denominator is 10, 100, 1000, or some other power of 10.

Here's an example: We'll start with three quarters. What is the first number in the 4 "times table" that you can multiply and get a multiple of 10? Can you multiply 4 by something to get 10? No. Can you multiply it by something to get 100? Yes! 4 X 25 is 100.

So multiply the numerator by 25, which is 75 over 100

We know fractions are really a division problem, and we also know that dividing by 100, means we move the decimal 2 places to the left.

So, 75 over 100 = .75

Lets try another example - Convert one fifth to a decimal.

First find a power of 10 that 5 goes into evenly, which is 2.

Multiply the numerator and denominator by 2, which is two tenths.

Dividing 2 by 10 means we move the decimal place 1 place to the left.

So 1/5 = 0.5

Converting Fractions to Percent

Here is a quick method to convert fraction to percent and a strategy for answering on a multiple choice test that will save you valuable exam time.

First, remember that a fraction is a division problem: you're dividing the bottom number into the top.

Taking an example, convert 2/3 into percent.

The first method is to multiple the numerator by 100 and divide. So,

(2 X 100) / 2 = 100/3 = 66.66

Add a % sign and you have the answer, 66.66%

If you're doing these conversions on a multiple-choice test, here's an idea that might be even easier and faster. Let's say you have a fraction of 1/8 and you're asked to convert to percent.

Since we know that "percent" means hundredths, ask yourself what number we can multiply 8 by to get 100. Since there is no number, ask what number gets us close to 100.

— Fractions, Decimals and Percent —

That number is 12: 8 X 12 = 96. So it gets us a little less than 100. Now, whatever you do to the denominator, you have to do to the numerator. Let's multiply 1 X 12 and we get 12. However, since 96 is a little less than 100, we know that our answer will be a little MORE than 12%.

Look at the choices and eliminate the obvious wrong choices. So if your possible answers on the multiple-choice test are these:

a) 8.5% b) 19% c) 12.5% d) 25%

then we know the answer is c) 12.5%, because it's a little MORE than the 12 we got in our math problem above. Here all the choices except choice C 12.5% can be eliminated.

You don't have to know the exact correct answer, just enough to estimate, then eliminate the obviously wrong answers.

This was an easy example to demonstrate the strategy, but don't be fooled! You probably won't get such an easy question on your exam. By estimating your answer quickly, then eliminating obviously incorrect choices immediately, you save precious exam time.

Decimal Tips, Tricks and Short-cuts

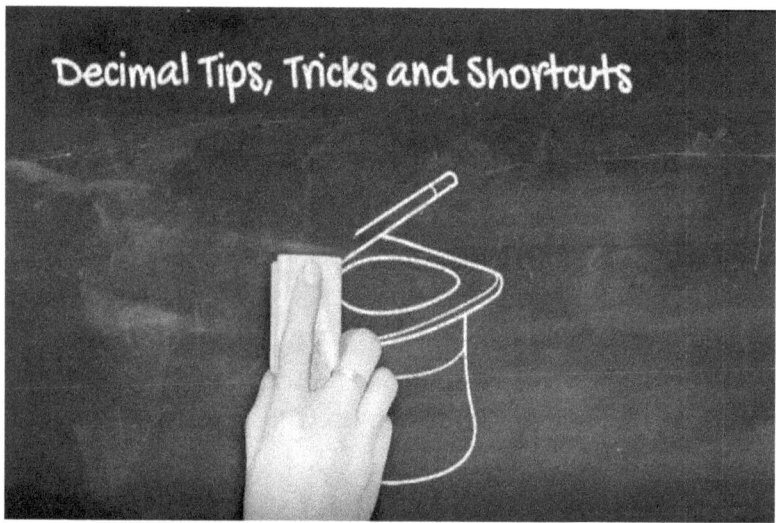

HTTPS://YOUTU.BE/4KTB4CK1SO8

Converting Decimals to Fractions

Converting decimals to fractions is easy if you say it the right way! If you say "point one" or "point 25," you'll have trouble.

But if you say, "one tenth" and "twenty-five hundredths," then you have already solved it! That's because, if you know your fractions, you know that "one tenth" looks like this: 1/10. And "twenty-five hundredths" looks like this: 25/100.

Even if you have digits before the decimal, such as

3.4, learning how to say the word will help you with the conversion into a fraction. It's not "three point four," it's "three and four tenths." Knowing this, you know that the fraction which looks like "three and four tenths" is 3 4/10.

The conversion is not complete until you reduce the fraction to its lowest terms: It's not 25/100, but 1/4.

Converting Decimals to Percent

Changing a decimal to a percent is easy if you remember one thing: multiply by 100.

For example, if you start with .45, simply multiply it by 100 for 45. Then add the % sign to the end - 45%.

Think of it this way: take out the decimal point, add a percent sign on the opposite side. In other words, the decimal on the left is replaced by the % on the right.

It doesn't work quite that easily if the decimal is in the middle of the number. For example, 3.7. Here, take out the decimal in the middle and replace it with a 0 % at the end. So 3.7 converted to decimal is 370%.

Percent Tips, Tricks and Short-cuts

Percent problems are not nearly as scary as they appear, if you remember this neat trick:

Draw a cross as in:

Portion	Percent
Whole	100

In the upper left, write PORTION. In the bottom left write WHOLE. In the top right, write PERCENT and in the bottom right, write 100. Whatever your problem is, you will leave blank the unknown, and fill in the other four parts. For example, let's suppose your problem is: Find 10% of 50. Since we know the 10% part, we put 10 in the percent corner. Since the whole number in our problem is 50, we put that in the corner marked whole. You always put 100 underneath the percent, so we leave it as is, which leaves only the top left corner blank. This is where we'll put our answer. Now simply multiply the two corner numbers that are NOT 100. Here, it's 10 X 50. That gives us 500. Now divide this by the remaining corner, or 100, to get a final answer of 5. 5 is the number that goes in the upper-left corner, and is your final solution.

Another hint to remember: Percents are the same thing as hundredths in decimals. So .45 is the same as 45 hundredths or 45 percent.

Converting Percents to Decimals

Percents are just a type of decimal, so it should be no surprise that converting between the two is actually fairly simple. Here are a few tricks and Short-cuts to keep in mind:

- Remember that percent literally means "per 100" or "for every 100." So when you speak of 30% you're saying 30 for every 100 or the fraction 30/100. In basic math, you learned that fractions that have 10 or 100 as the denominator can easily be turned to a decimal. 30/100 is thirty hundredths, or expressed as a decimal, .30.
- Another way to look at it: To convert a percent to a decimal, simply divide the number by 100. So for instance, if the percent is 47%, divide 47 by 100.

― Fractions, Decimals and Percent ―

The result will be .47. Get rid of the % mark and you're done.
- Remember that the easiest way of dividing by 100 is by moving your decimal two spots to the left.

Converting Percents to Fractions

Converting percents to fractions is easy. After all, a percent is just a type of fraction; it tells you what part of 100 that you're talking about. Here are some simple ideas for making the conversion from a percent to a fraction:

- If the percent is a whole number -- say 34% -- then simply write a fraction with 100 as the denominator (the bottom number). Then put the percentage itself on top. So 34% becomes 34/100.
- Now reduce as you would reduce any percent. In this case, by dividing 2 into 34 and 2 into 100, you get 17/50.
- If your percent is not a whole number -- say 3.4% --then convert it to a decimal expressed as hundredths. 3.4 is the same as 3.40 (or 3 and forty hundredths). Now ask yourself how you would express "three and forty hundredths" as a fraction. It would, of course, be 3 40/100. Reduce this and it becomes 3 2/5.

Most Common Basic Math Mistakes

Careless Mistakes. Misreading a problem, simple arithmetic mistakes, or other careless errors

Not showing all of the steps. This makes it difficult for the teacher to understand how you got the answer.

Not checking your work or not reviewing their test before turning it in. Never leave the test room early!

Not understanding the problem and solving it with the wrong method.

Not understanding of basic concepts and operations, such as fractions, decimals, and basic algebra.

Not paying attention to the units of measure. Not understanding basic terminology, such as "factor," "product," and "quotient."

Not paying attention to the sign of the answer or confusing the sign.

Not using the correct formula or equation for the problem.

Fractions, Decimals and Percent

Answer Sheet

	A	B	C	D	E		A	B	C	D	E
1	○	○	○	○	○	21	○	○	○	○	○
2	○	○	○	○	○	22	○	○	○	○	○
3	○	○	○	○	○	23	○	○	○	○	○
4	○	○	○	○	○	24	○	○	○	○	○
5	○	○	○	○	○	25	○	○	○	○	○
6	○	○	○	○	○	26	○	○	○	○	○
7	○	○	○	○	○	27	○	○	○	○	○
8	○	○	○	○	○	28	○	○	○	○	○
9	○	○	○	○	○	29	○	○	○	○	○
10	○	○	○	○	○	30	○	○	○	○	○
11	○	○	○	○	○						
12	○	○	○	○	○						
13	○	○	○	○	○						
14	○	○	○	○	○						
15	○	○	○	○	○						
16	○	○	○	○	○						
17	○	○	○	○	○						
18	○	○	○	○	○						
19	○	○	○	○	○						
20	○	○	○	○	○						

Practice Questions

1. 2/3 + 5/12 =

    a. 9/17`
    b. 3/11
    c. 7/12
    d. 1 1/12

2. 3/5 + 7/10 =

    a. 1 1/10
    b. 7/10
    c. 1 3/10
    d. 1 1/12

3. 4/5 − 2/3 =

    a. 2/2
    b. 2/13
    c. 1
    d. 2/15

4. 13/16 − 1/4 =

    a. 1
    b. 12/12
    c. 9/16
    d. 7/16

— Fractions, Decimals and Percent —

5. 15/16 x 8/9 =

    a. 5/6
    b. 16/37
    c. 2/11
    d. 5/7

6. 3/4 x 5/11 =

    a. 2/15
    b. 15/44
    c. 3/19
    d. 15/44

7. 5/8 ÷ 2/3 =

    a. 15/16
    b. 10/24
    c. 5/12
    d. 1 2/5

8. 2/15 ÷ 4/5 =

    a. 6/65
    b. 6/75
    c. 5/12
    d. 1/6

9. In a class of 83 students, 72 are present. What percent of the students are absent? Provide answer up to two significant digits.

    a. 12%
    b. 13%
    c. 14%
    d. 15%

10. A woman spent 15% of her income on an item and ends with $120. What percentage of her income is left?

    a. 12%
    b. 85%
    c. 75%
    d. 95%

11. X% of 120 = 30. Solve for X.

    a. 15
    b. 12
    c. 4
    d. 25

12. Simplify 6 3/5 – 4 4/5

    a. 1 4/5
    b. 2 3/5
    c. 2 9/5
    d. 1 1/5

— Fractions, Decimals and Percent —

13. Express 25% as a fraction.

   a. 1/4
   b. 7/40
   c. 6/25
   d. 8/28

14. Express 125% as a decimal.

   a. .125
   b. 12.5
   c. 1.25
   d. 125

15. Express 24/56 as a reduced common fraction.

   a. 4/9
   b. 4/11
   c. 3/7
   d. 3/8

16. Express 71/1000 as a decimal.

   a. .71
   b. .0071
   c. .071
   d. 7.1

17. .4% of 36 is

   a. 1.44
   b. .144
   c. 14.4
   d. 144

18. Express 0.27 + 0.33 as a fraction.

    a. 3/6

    b. 4/7

    c. 3/5

    d. 2/7

19. 8 is what percent of 40?

    a. 10%

    b. 15%

    c. 20%

    d. 25%

20. 3.14 + 2.73 + 23.7 =

    a. 28.57

    b. 30.57

    c. 29.56

    d. 29.57

21. What is 1/3 of 3/4?

    a. 1/4

    b. 1/3

    c. 2/3

    d. 3/48

22. 15 is what percent of 200?

    a. 7.5%

    b. 15%

    c. 20%

    d. 17.50%

— Fractions, Decimals and Percent —

23. A boy has 5 red balls, 3 white balls and 2 yellow balls. What percent of the balls are yellow?

a. 2%
b. 8%
c. 20%
d. 12%

24. Add 10% of 300 to 50% of 20

a. 50
b. 40
c. 60
d. 45

25. Convert 75% to a fraction.

a. 2/100
b. 85/100
c. 3/4
d. 4/7

26. Multiply 3 by 25% of 40.

a. 75
b. 30
c. 68
d. 35

27. What is 10% of 30 multiplied by 75% of 200?

    a. 450

    b. 750

    c. 20

    d. 45

28. Convert 4/20 to percent.

    a. 25%

    b. 20%

    c. 40%

    d. 30%

29. Write 765.3682 to the nearest 1000th.

    a. 765.368

    b. 765.36

    c. 765.3682

    d. 765.3

30. What number is in the ten thousandths place in 1.7389?

    a. 1

    b. 8

    c. 9

    d. 3

--- Fractions, Decimals and Percent ---

Answer Key

1. D
A common denominator is needed, which both 3 and 12 will divide into. So, 8 + 5/12 = 13/12 = 1 1/12

2. C
A common denominator is needed for 5 and 10.
6 + 7/10 = 13/10 = 1 3/10

3. D
A common denominator is needed for 5 and 3.
12 - 10/15 = 2/15

4. C
A common denominator is needed for 16 and 4.
13 - 4/16 = 9/16

5. A
Since there are common numerators and denominators to cancel out, we cancel out 15/16 x 8/9 to get 5/2 x 1/3, and then multiply numerators and denominators to get 5/6

6. D
Since there are no common numerators and denominators to cancel out, we simply multiply the numerators and then the denominators. So 3 x 5/4 x 11 = 15/44

7. A
To divide fractions, multiply the first fraction with the inverse of the second. 5/8 x 3/2, = 15/16

8. D
Multiply the first fraction with the inverse of the second.
2/15 x 5/4, (cancel out) = 1/3 x 1/2 = 1/6

9. B
Number of absent students = 83 – 72 = 11
Percentage of absent students is found by proportioning the number of absent students to the total number of students in the class = 11 * 100/83 = 13.25

Checking the answer, round 13.25 to the nearest whole number: 13%.

10. B
Spent 15%, so 100% - 15% = 85%

11. D
X% of 120 = 30,
X/100 = 30/120
So X = 30/120 x 100/1
3000/120 = 300/12
X = 25

12. A
(6-4) (3/5 – 4/5) = 2 (3-4/5) = since 3 is less than 4, we would have to subtract 1 from the whole number besides the fraction, therefore 1 4/5

13. A
25% = 25/100 = 1/4

14. C
125/100 = 1.25

15. C
24/56 = 3/7 (divide numerator and denominator by 8)

16. C
Converting a fraction into a decimal – divide the numerator by the denominator – so 71/1000 = .071. Dividing by 1000 moves the decimal point 3 places to the left.

Fractions, Decimals and Percent

17. B
.4/100 * 36 = .4 * 36/100 = .144

18. C
To convert a decimal to a fraction, take the places of decimal as your denominator, here, 2, so in 0.27, '7' is in the 100th place, so the fraction is 27/100 and 0.33 becomes 33/100.

Next estimate the answer quickly to eliminate obvious wrong choices. 27/100 is about 1/4 and 33/100 is 1/3. 1/3 is slightly larger than 1/4, and 1/4 + 1/4 is 1/2, so the answer will be slightly larger than 1/2.
Looking at the choices, Choice A can be eliminated since 3/6 = 1/2. Choice D, 2/7 is less than 1/2 and be eliminated. The answer is going to be Choice B or Choice C.

Do the calculation, 0.27 + 0.33 = 0.60 and 0.60 = 60/100 = 3/5, Choice C is correct.

19. C
This is an easy question, and shows how you can solve some questions without doing the calculations. The question is, 8 is what percent of 40. Take easy percentages for an approximate answer and see what you get.

10% is easy to calculate because you can drop the zero, or move the decimal point. 10% of 40 = 4, and 8 = 2 X 4, so, 8 must be 2 X 10% = 20%.

Here are the calculations which confirm the quick approximation.
8/40 = X/100 = 8 * 100 / 40X = 800/40 = X = 20

20. D
3.14 + 2.73 = 5.87 and 5.87 + 23.7 = 29.57

21. A
1/3 X 3/4 = 3/12 = 1/4
To multiply fractions, multiply the numerator and denominator.

22. A
15/200 = X/100
200X = (15 * 100)
1500/200 Cancel zeros in the numerator and denominator
15/2 = 7.5%.

Notice that the questions asks, What 15 is what percent of 200? The question does *not* ask, what is 15% of 200! The answers are very different.

23. C
Total no. of balls = 10, no. of yellow balls = 2, answer = 2/10 X 100 = 20%.

24. B
10% of 300 = 30 and 50% of 20 = 10 so 30 + 10 = 40.

25. C
75% = 75/100 = 3/4

26. B
25% of 40 = 10 and 10 x 3 = 30

27. A
10% of 30 = 3 and 75% of 200 = 150, 3 X 150 = 450
28. B
4/20 X 100 = 1/5 X 100 = 20%

29. A
The number is 51.738. The last digit, in the 1,000th place, 2, is less than 5, so it is discarded. Answer = 765.368.

30. C
9 is in the ten thousandths place in 1.7389, which is 4 places to the right of the decimal point.

Arithmetic, Number Manipulation and Non-Geometric Comparison

Number Manipulation Strategy

Here is an example of a number manipulation question:

What number divided by 5 is 1/4 of 100?

    a. 125
    b. 150
    c. 75
    d. 225

The answer is A – 125.

Number manipulation questions are 'double barreled' questions, asking you to make 2 calculations – calculate 'something' of 'something else.' This increases the chance

— Arithmetic —

of error and forces you to think very carefully.

The best strategy for most math questions is to break them down into little pieces, solve the pieces, and then put them back together. Here is a systematic strategy for solving number manipulation problems.

Example #1

Step 1 – Break it down.

> What number divided by 5

is

> 1/4 of 100?

Taking the example above, there are 2 parts:

Part 1 - A number, call it Z, divided by 5, so X/5
Part 2 – 1/4 of 100 = 25

The two parts are connected by 'is', which mean 'equals,' so we can re-write the problem:

Step 2 – Re-Write the problem

X/5 = 25

Step 3 - Solve
Then solve for X,

X = 5.

Example #2

What number subtracted from 50 is 1/5 of 40?

This example has 'subtracted from' instead of 'is.' The operator phrase, 'subtracted from' can also be, 'multiplied by,' 'added to,' or similar.

Step 1 – Break it down.

> What number subtracted from 50

is

> 1/5 of 40?

Step 2 – Re-Write the problem

Let Z be the number, so we can write as:

50 – Z is 1/5 of 40   or

50 – Z = 1/5 of 40

Step 3 - Solve the Problem

50 – Z = 40/5
50 – Z = 8
Z = 42

Arithmetic

Answer Sheet

1. Ⓐ Ⓑ Ⓒ Ⓓ   21. Ⓐ Ⓑ Ⓒ Ⓓ   41. Ⓐ Ⓑ Ⓒ Ⓓ   61. Ⓐ Ⓑ Ⓒ Ⓓ
2. Ⓐ Ⓑ Ⓒ Ⓓ   22. Ⓐ Ⓑ Ⓒ Ⓓ   42. Ⓐ Ⓑ Ⓒ Ⓓ   62. Ⓐ Ⓑ Ⓒ Ⓓ
3. Ⓐ Ⓑ Ⓒ Ⓓ   23. Ⓐ Ⓑ Ⓒ Ⓓ   43. Ⓐ Ⓑ Ⓒ Ⓓ   63. Ⓐ Ⓑ Ⓒ Ⓓ
4. Ⓐ Ⓑ Ⓒ Ⓓ   24. Ⓐ Ⓑ Ⓒ Ⓓ   44. Ⓐ Ⓑ Ⓒ Ⓓ   64. Ⓐ Ⓑ Ⓒ Ⓓ
5. Ⓐ Ⓑ Ⓒ Ⓓ   25. Ⓐ Ⓑ Ⓒ Ⓓ   45. Ⓐ Ⓑ Ⓒ Ⓓ   65. Ⓐ Ⓑ Ⓒ Ⓓ
6. Ⓐ Ⓑ Ⓒ Ⓓ   26. Ⓐ Ⓑ Ⓒ Ⓓ   46. Ⓐ Ⓑ Ⓒ Ⓓ   66. Ⓐ Ⓑ Ⓒ Ⓓ
7. Ⓐ Ⓑ Ⓒ Ⓓ   27. Ⓐ Ⓑ Ⓒ Ⓓ   47. Ⓐ Ⓑ Ⓒ Ⓓ   67. Ⓐ Ⓑ Ⓒ Ⓓ
8. Ⓐ Ⓑ Ⓒ Ⓓ   28. Ⓐ Ⓑ Ⓒ Ⓓ   48. Ⓐ Ⓑ Ⓒ Ⓓ   68. Ⓐ Ⓑ Ⓒ Ⓓ
9. Ⓐ Ⓑ Ⓒ Ⓓ   29. Ⓐ Ⓑ Ⓒ Ⓓ   49. Ⓐ Ⓑ Ⓒ Ⓓ   69. Ⓐ Ⓑ Ⓒ Ⓓ
10. Ⓐ Ⓑ Ⓒ Ⓓ   30. Ⓐ Ⓑ Ⓒ Ⓓ   50. Ⓐ Ⓑ Ⓒ Ⓓ   70. Ⓐ Ⓑ Ⓒ Ⓓ
11. Ⓐ Ⓑ Ⓒ Ⓓ   31. Ⓐ Ⓑ Ⓒ Ⓓ   51. Ⓐ Ⓑ Ⓒ Ⓓ   71. Ⓐ Ⓑ Ⓒ Ⓓ
12. Ⓐ Ⓑ Ⓒ Ⓓ   32. Ⓐ Ⓑ Ⓒ Ⓓ   52. Ⓐ Ⓑ Ⓒ Ⓓ   72. Ⓐ Ⓑ Ⓒ Ⓓ
13. Ⓐ Ⓑ Ⓒ Ⓓ   33. Ⓐ Ⓑ Ⓒ Ⓓ   53. Ⓐ Ⓑ Ⓒ Ⓓ   73. Ⓐ Ⓑ Ⓒ Ⓓ
14. Ⓐ Ⓑ Ⓒ Ⓓ   34. Ⓐ Ⓑ Ⓒ Ⓓ   54. Ⓐ Ⓑ Ⓒ Ⓓ   74. Ⓐ Ⓑ Ⓒ Ⓓ
15. Ⓐ Ⓑ Ⓒ Ⓓ   35. Ⓐ Ⓑ Ⓒ Ⓓ   55. Ⓐ Ⓑ Ⓒ Ⓓ   75. Ⓐ Ⓑ Ⓒ Ⓓ
16. Ⓐ Ⓑ Ⓒ Ⓓ   36. Ⓐ Ⓑ Ⓒ Ⓓ   56. Ⓐ Ⓑ Ⓒ Ⓓ   76. Ⓐ Ⓑ Ⓒ Ⓓ
17. Ⓐ Ⓑ Ⓒ Ⓓ   37. Ⓐ Ⓑ Ⓒ Ⓓ   57. Ⓐ Ⓑ Ⓒ Ⓓ   77. Ⓐ Ⓑ Ⓒ Ⓓ
18. Ⓐ Ⓑ Ⓒ Ⓓ   38. Ⓐ Ⓑ Ⓒ Ⓓ   58. Ⓐ Ⓑ Ⓒ Ⓓ   78. Ⓐ Ⓑ Ⓒ Ⓓ
19. Ⓐ Ⓑ Ⓒ Ⓓ   39. Ⓐ Ⓑ Ⓒ Ⓓ   59. Ⓐ Ⓑ Ⓒ Ⓓ   79. Ⓐ Ⓑ Ⓒ Ⓓ
20. Ⓐ Ⓑ Ⓒ Ⓓ   40. Ⓐ Ⓑ Ⓒ Ⓓ   60. Ⓐ Ⓑ Ⓒ Ⓓ   80. Ⓐ Ⓑ Ⓒ Ⓓ

Practice Questions

1. 389 + 454 =

    a. 853
    b. 833
    c. 843
    d. 863

2. 9,177 + 7,204 =

    a. 16,4712
    b. 16,371
    c. 16,381
    d. 15,412

3. 8,390 - 5,239 =

    a. 3,261
    b. 3,151
    c. 3,161
    d. 3,101

4. 643 - 587 =

    a. 56
    b. 66
    c. 46
    d. 55

— Arithmetic —

5. 3,406 - 2,767 =

    a. 629

    b. 720

    c. 639

    d. 649

6. 149 × 7 =

    a. 1032

    b. 1043

    c. 1059

    d. 10

7. 309 × 17 =

    a. 5,303

    b. 4,913

    c. 4,773

    d. 5,253

8. 491 ÷ 9 =

    a. 54 r5

    b. 56 r6

    c. 57 r5

    d. 51 r

9. What is the best unit to measure the weight of a tanker truck?

    a. tons

    b. ounces

    c. quarts

    d. feet

10. What is the best unit to measure the length of a needle?

   a. centimeters
   b. feet
   c. ounces
   d. meters

11. What is the best unit to measure the depth of a lake?

   a. meters
   b. tons
   c. inches
   d. pounds

12. Jim bought $15.34 worth of candy and his sister bought $12.57 worth. How much less did his sister buy?

   a. 3.49
   b. 2.77
   c. 2.27
   d. 1.77

13. Andrea earned $877.20 on her summer vacation and Bill earned 107.82 less. How much did he earn?

   a. 107.48
   b. 459.44
   c. 789.38
   d. 769.38

14. 20% of the students in a class of 50 leave early. Half of the remaining students stay after class. The rest leave at the bell. How many students stay after class?

    a. 10
    b. 20
    c. 15
    d. 12

15. Which of the following is true?

    a. 1/3 > 1/4 > 1/5
    b. 1/5 > 1/3 > 1/4
    c. 1/4 > 1/5 > 1/3
    d. 1/5 > 1/4 > 1/3

16. Which of the following is true?

    a. 2/3 < 3/4 > 1/3
    b. 1/3 > 2/3 > 3/4
    c. 3/4 > 2/3 < 1/3
    d. 3/4 < 1/3 > 2/3

17. Which of the following is between 0.51 and 0.98?

    a. 0.75
    b. 0.25
    c. 0.45
    d. 0.99

18. Which of the following is between 0.0067 and 0.34?

    a. 0.078

    b. 0.42

    c. 0.0005

    c. 0.12

19. You are asked to calculate 40% of 250 and use the formula:

40/100 = X/250. What other equation could you use?

    a. 40X = 25000

    b. (40 * 250)/100

    c. (40 * 100)/250

    d. 100X = (40 * 250)

20. You are asked to calculate 1/3 of 150. Which of the following equations is correct?

    a. 3/150

    b. 150/3

    c. 150 * 3

    d. None of the above

21. Barry rode the ferry to work and counted 5 whales and 46 seagulls. On the way home, he counted 1 whale and 23 seagulls. How many fewer seagulls did he see on the way back?

    a. 25%

    b. 50%

    c. 45%

    d. 65%

—— Arithmetic ——

22. -75 + (-14) =

  a. -99
  b. -89
  c. 89
  d. 99

23. 50 - (-22) =

  a. 78
  b. -28
  c. -72
  d. 72

Non-Geometric Comparison

24. Examine the following and find the best answer.

  1. (3 X 4) - 7
  2. (3 X 4) - 3
  3. (3 X 5) – 10

  a. 1 and 3 are equal
  b. 1 and 3 are not equal
  c. 1 and 2 are equal
  d. 2 and 3 are equal

25. Examine the following and find the best answer.

1. (5 X 7) + 3
2. (5 X 6) + 9
3. (5 X 5) + 18

    a. 1 is less than 3 and greater than 2
    b. 3 is greater than 1 and 2
    c. 1 is less than 2 and 3
    d. 2 is greater than 3 and 1

26. Examine the following and find the best answer.

1. 23 - 15
2. 32 - 23
3. 42 - 26

    a. $1 < 2 < 3$
    b. $2 > 3 > 1$
    c. $3 > 1 > 2$
    d. $1 < 2 > 3$

27. Examine the following and find the best answer.

1. 4^2
2. 2^4
3. 2^3
4. 3^2

    a. 1 and 3 are equal
    b. 2 and 3 are equal
    c. 3 and 4 are equal
    d. None of the above

— Arithmetic —

28. Examine the following and find the best answer.

A. (2 X 40) - 20
B. (3 X 14) + 15
C. (3 X 15) – 8

    a. A < B < C
    b. A > B > C
    c. B > C < A
    d. C > B < A

29. What is 6 more than 1/3 of 25?

    a. 8.33
    b. 15
    c. 14.33
    d. 18

30. What number is the largest?

    a. 50% of 50
    b. 2/5 of 200
    c. 0.2 portion of 20
    d. 1/5th of 30

31. What is the smallest value?

    a. 0.4 portion of 200
    b. 50% of 100
    c. 0.06 portion of 2000
    d. 2% of 1000

32. What number is the largest?

   a. 5 % of 400
   b. 25% of 4000
   c. 2 % of 500
   d. 8% of 1000

33. Examine the following and find the best answer.

1. (4 X 7) - 12
2. (2 X 14) + 5
3. (3 X 10) – 8

   a. 1 < 2 < 3
   b. 1 < 2 > 3
   c. 2 > 3 < 1
   d. 3 > 2 < 1

34. Examine the following and find the best answer.

1. (8 X 2) X (2 X 3)
2. (4 X 8) + (2 X 6)
3. (3 X 4) X (9 - 2)
4. (3 X 7) - (2 X 6)

   a. 1 < 3 < 4
   b. 1 > 4 < 3
   c. 2 > 1 > 3
   d. 4 < 1 < 3

35. Examine the following and find the best answer.

1. 12 + (15 - 4)
2. 9 + (24 - 2)
3. 2 + (19 + 7)
4. 15 + (26 - 19)

    a. 1 < 3 < 4
    b. 1 < 4 < 3
    c. 2 > 1 < 3
    d. 4 > 1 < 3

36. Examine the following and find the best answer.

1. (23 - 7) * 2
2. (9 + 12) / 7
3. (13 - 7) X 2
4. (17 + 4) + 1

    a. 2 is the smallest
    b. 4 is the largest
    c. 1 is the smallest
    d. None of the Above

37. Examine the following and find the best answer.

1. 1 + (6 * 7)
2. 7 - 22
3. 9 + (13 X 2)
4. 6 * (5 + 2)

    a. 3 and 4 are greater than 1
    b. 3 and 2 are greater than 1
    c. 3 is less than 2
    d. None of the Above

38. Examine the following and find the best answer.

1. 19 + (4 X 5)
2. 13 - (4 + 6)
3. 7 + (3 X 7)

    a. 1 > 2 < 3
    b. 2 > 3 < 1
    c. 2 < 3 > 1
    d. 1 < 2 < 3

39. Examine the following and find the best answer.

1. 18 + 7
2. 4 X 8
3. 3 X 5
4. 2 X 10

    a. (#2 - #4) > #1
    b. (#4 - #3) > #2
    c. (#2 - #1) < #3
    d. (#1 + #2) > #4

40. Examine the following and find the best answer.

1. (6 X 3) X (3 X 4)
2. (4 X 5) + (3 X 6)
3. (3 X 4) X (5 - 2)

    a. 3 is the largest
    b. 1 is less than 2
    c. 3 is greater than 1
    d. None of the Above

Arithmetic

41. Examine the following and find the best answer.

1. (6 X 1) X (2 X 4)
2. (4 X 7) + (5 X 6)
3. (3 X 4) X (9 - 2)

    a. 2 is the largest
    b. 3 is greater than 1
    c. 1 is greater than 2
    d. None of the Above.

42. Examine the following and find the best answer.

1. 6 X (4 + 3)
2. 4 X (5 X 3)
3. 7 + (3 X 6)

    a. 1 < 2 < 3
    b. 1 < 2 > 3
    c. 2 > 3 > 1
    d. 2 < 3 > 1

43. Examine the following and find the best answer.

1. 50 - (5 X 4)
2. 100 - (6 X 6)
3. 45 - (4 X 2)

    a. 1 < 3 < 2
    b. 1 > 3 > 2
    c. 2 > 1 > 3
    d. 2 < 1 > 3

44. Examine the following and find the best answer.

1. 35 - (6 X 3)
2. 100 - (6 X 10)
3. 75 - (4 X 7)

    a. 1 < 3 < 2
    b. 1 > 3 > 2
    c. 2 > 1 > 3
    d. 2 < 1 > 3

45. Examine the following and find the best answer.

1. 4 X (2 + 3)
2. 5 X (7 X 3)
3. 9 + (3 X 7)

    a. 3 < 2 > 1
    b. 1 > 2 < 3
    c. 2 > 3 < 1
    d. 2 < 3 < 1

46. Examine the following and find the best answer.

(4 X 3) X (3 X 7)
(7 X 5) + (2 X 5)
(4 X 3) X (7 - 2)

    a. 1 > 2 > 3
    b. 2 > 1 > 3
    c. 3 < 2 < 1
    d. 2 < 1 < 3

— Arithmetic —

47. Examine the following and find the best answer.

1. 12 + (5 X 3)
2. 15 - (9 + 4)
3. 4 + (2 X 9)

    a. 1 > 2 > 3
    b. 2 < 1 > 3
    c. 3 < 2 < 1
    d. 2 < 1 < 3

48. Examine the following and find the best answer.

1. 15 X 6
2. 37 X 5
3. 24 X 2

    a. 1 = 3
    b. 1 = 2
    c. 1 > 2 > 3
    d. 3 < 1 < 2

49. Examine the following and find the best answer.

1. (6 X 4) - 5
2. (2 X 8) - 3
3. (3 X 5) – 8

    a. 1 and 3 are equal
    b. 1 and 3 are not equal
    c. 1 and 2 are equal
    d. 2 and 3 are equal

Number Manipulation

50. 2/3 of what number added to 10 is 3 times 15?

    a. 50

    b. 52.5

    c. 65

    d. 72.8

51. What number is 25 more than 1/3 of 27?

    a. 34

    b. -16

    c. 20

    d. 18

52. What number is 10 less than 5 squared?

    a. 25

    b. 10

    c. 15

    d. 40

53. What is 12 more than 3/4 of 40?

    a. 42

    b. 35

    c. 52

    d. 18

— Arithmetic —

54. What is 25 more than 30% of 120?

    a. 75
    b. 50
    c. 65
    d. 35

55. What number subtracted from 75 leaves 15 more than 3/5 of 75?

    a. 25
    b. 15
    c. 30
    d. 12

56. What number divided by 5 is 2/3 of 100?

    a. 333.33
    b. 444.44
    c. 250
    d. 100

57. 1/2 of what number is 9 times 5?

    a. 60
    b. 120
    c. 90
    d. 45

58. What number multiplied by 8 is 10 less than 58?

    a. 5
    b. 6
    c. 7
    d. 8

59. What number subtracted from 25 is 1/5 of 40?

   a. 22
   b. 17
   c. 34
   d. 12

60. What number is 1 less than 1/4 of 16?

   a. 1
   b. 2
   c. 3
   d. 4

61. 1/2 of what number is 8 times 7?

   a. 60
   b. 120
   c. 112
   d. 45

62. What number multiplied by 10 is 15 less than 65?

   a. 5
   b. 6
   c. 7
   d. 8

63. What number subtracted from 25 is 1/5 of 60?

   a. 22
   b. 13
   c. 34
   d. 12

―――― Arithmetic ――――

64. What is the smallest value?

a. 4 % of 4
b. 5 % of 5
c. 0.3% of 12
d. 1/5 of 100

65. What is 8 more than 2/5 of 20?

a. 10
b. 12
c. 16
d. 8

66. What is 6 more than 50% of 50?

a. 31
b. 41
c. 25
d. 26

67. What number subtracted from 100 leaves 7 more than 3/4 of 40?

a. 50
b. 63
c. 47
d. 75

68. What number divided by 5 is 1/4 of 100?

a. 125
b. 150
c. 75
d. 225

69. 1/5 of what number is 5 times 10?

    a. 150
    b. 200
    c. 250
    d. 100

70. What number multiplied by 5 is 10 less than 52?

    a. 8.4
    b. 10.24
    c. 20
    d. 22.5

71. What number subtracted from 25 is 1/5 of 20?

    a. 125
    b. 150
    c. 50
    d. 21

72. What number is 15 less than 3/5 of 40?

    a. 42
    b. 17
    c. 9
    d. 18

73. What number is 10 times 1/2 of 60?

    a. 100
    b. 300
    c. 250
    d. 75

— Arithmetic —

74. 1/4 of what number is 8 times 10?

   a. 150
   b. 200
   c. 320
   d. 100

75. 1/2 of what number is 4 X 15

   a. 120
   b. 40
   c. 50
   d. 60

Answer Key

1. C
389 + 454 = 843

2. C
9,177 + 7,204 = 16,381

3. B
8,390 - 5,239 = 3,151

4. A
643 - 587 = 56

5. C
3,406 - 2,767 = 639

6. B
149 × 7 = 1043

7. D
309 × 17 = 52,53

8. A
491 ÷ 9 = 54 r5

9. A
Tons are the best units to measure the weight of a tanker truck.

10. A
Centimeters are the best measure for the length of a needle.

11. A
Meters are the best measures for the depth of a lake.

— Arithmetic —

12. B
She bought $2.77 less candy. 15.34 - 12.57 = 2.77

13. D
He earned 769.38 less than Andrea. 877.20 - 107.82 = 769.38.

14. B
20% of 50 = 10 students. Half of the remaining stay after class, 50 - 10 = 40 / 2 = 20.

15. A
1/3 > 1/4 > 1/5

16. A
2/3 < 3/4 > 1/3

17. A
0.75 is between 0.51 and 0.98.

18. A
0.078 is between 0.0067 and 0.34.

19. B
Another equation to calculate 40% of 250 is (40 * 250)/100.

20. B
The correct formula to calculate 1/3 of 150 is 150/3.

21. B
There were 46 on the way there and 23 on the way back.

46/23 = 50%

22. B
-75 + (-14) = -89

23. A
50 - (-22) = 78

Non-Geometric Comparisons

24. A
#1 = 5
#2 = 9
#3 = 5
1 and 3 are equal.

25. C
#1 = 38
#2 = 39
#3 = 43
1 is less than 2 and less than 3.

26. A
#1 = 8
#2 = 9
#3 = 16
1 < 2 < 3

27. D
#1 = 16
#2 = 16
#3 = 8
#4 = 9
None of the Above.

28. B
A = 60
B = 57
C = 37
A > B > C

29. C
1/3 of 25 = 8.33 + 6 = 14.33

―――― Arithmetic ――――

30. B
a. 50% of 50 = 25
b. 2/5 of 200 = 80
c. 0.2 portion of 20 = 4
d. 1/5 of 30 = 6
B is the largest.

31. D
a. 0.4 portion of 200 = 80
b. 50% of 100 = 50
c. 0.06 portion of 2000 = 120
d. 2% of 1000 = 20
D is the smallest

32. B
a. 5% of 400 = 20
b. 25% of 4000 = 1000
c. 2% of 500 = 10
d. 8% of 1000 = 80

33. B
#1 = 16
#2 = 33
#3 = 22
1 < 2 > 3

34. B
1. 96
2. 44
3. 84
4. 9
1 > 4 < 3

35. C
#1 = 23
#2 = 31
#3 = 28
#4 = 22
2 > 1 < 3

36. A
#1 = 32
#2 = 3
#3 = 12
#4 = 22
2 is the smallest.

37. D
#1 = 43
#2 = 15
#3 = 35
#4 = 42
None of the Above.

38. A
#1 = 39
#2 = 3
#3 = 28
1 > 2 < 3

39. D
1. 25
2. 32
3. 15
4. 20
(#1 + #2) = 57 > 20

Arithmetic

40. D
#1 = 216
#2 = 360
#3 = 36
None of the Above.

41. B
#1 = 48
#2 = 58
#3 = 84
3 is greater than 1

42. B
#1 = 42
#2 = 60
#3 = 25
1 < 2 > 3

43. A
#1 = 30
#2 = 64
#3 = 37
1 < 3 < 2

44. B
#1 = 17
#2 = 40
#3 = 47
1 < 3 > 2

45. A
#1 = 20
#2 = 105
#3 = 30
3 < 2 > 1

46. B
#1 = 252
#2 = 350
#3 = 60
2 > 1 > 3

47. B
#1 = 27
#2 = 2
#3 = 22
2 < 1 > 3

48. D
#1 = 90
#2 = 185
#3 = 48
3 < 1 < 2

49. B
#1 = 19
#2 = 13
#3 = 7
1 and 3 are not equal

Number Manipulation

50. B
(2/3)Z + 10 = 3 X 15
2/3 Z = 45 − 10
Z = 35 X 3/2
Z = 52.5

51. A
X = 25 + (1/3 * 27)
X = 25 + 9
X = 34

— Arithmetic —

52. C
Z = (5 X 5) – 10
Z = 15

53. A
3/4 X 40 = 30 + 12 = 42

54. C
30% of 120 = 40 + 25 = 65

55. B
75 - x = 15 + (3/5 * 75)
75 - x = 45 + 15
75 - x = 60
x = 75 – 60 = 15

56. A
2/3 of 100 = 66.66 X 5 = 333.33

57. C
9 X 5 = 45 X 2 = 90

58. B
58 - 10 = 48 ÷ 8 = 6

59. B
1/5 of 40 = 8
25 – X = 8
X = 17

60. C
1/4 of 16 = 4
4 – 1 = 3

61. C
8 X 7 = 56 X 2 = 112

62. A
65 - 15 = 50 ÷ 10 = 5

63. B
1/5 of 60 = 12
25 – X = 12
X = 13

64. C
a. 4% of 4 = .16
b. 5% of 5 = 0.25
c. 0.3% of 12 = 0.036
d. 1/5 of 100 = 20
C is the smallest.

65. C
2/5 of 20 = 8 + 8 = 16

66. A
50% of 50 = 25 + 6 = 31

67. B
3/4 of 40 = 30 + 7 = 37
100 – X = 37
X = 63

68. A
x/5 = 1/4 * 100
x/5 = 25
x = 125

69. C
1/5X = 5 * 10
1/5X = 50
x = 250

— Arithmetic —

70. A
$5z = 52 - 10$
$Z = 10.4 - 2$
$Z = 8.4$

71. D
$25 - Z = 1/5 \times 20$
$25 - Z = 4$
$Z = 21$

72. C
$Z = (3/5 * 40) - 15$
$Z = 24 - 15$
$Z = 9$

73. B
$Z = 10 \times (1/2 \times 60)$
$Z = 10 \times 30$
$Z = 300$

74. C
$8 \times 10 = 80$
$1/4Z = 80$
$Z = 4 \times 80$
$Z = 320$

75. A
$4 \times 15 = 60$
$1/2\ x = 60$
$X = 120$

Basic Algebra

Solving One-Variable Linear Equations

Linear equations with variable x is an equation with the following form:
$$ax = b$$

where a and b are real numbers. If a=0 and b is different from 0, then the equation has no solution.

Let's solve one simple example of a linear equation with one variable:
$$4x - 2 = 2x + 6$$

When given this type of equation, move variables to one side, and real numbers to the other. Always remember: if you are changing sides, you are changing signs. Move all variables to the left, and real numbers to the right:
4x - 2 = 2x + 6
4x - 2x = 6 + 2
2x = 8
x = 8/2
x = 4

When 2x goes to the left it becomes -2x, and -2 goes to the right and becomes +2. After calculations, we find that x is 4, which is a solution of our linear equation.

Let's solve a little more complex linear equation:

2x - 6/4 + 4 = x
2x - 6 + 16 = 4x
2x - 4x = -16 + 6
-2x = -10
x = -10/-2
x = 5

We multiply whole equation by 4, to lose the fractional line. Now we have a simple linear equation. If we change sides, we change the signs.

Solving Two-Variable Linear Equations

If we have 2 or more linear equations with 2 or more variables, then we have a system of linear equations. The idea here is to express one variable using the other in one equation, and then use it in the second equation, so we get a linear equation with one variable. Here is an example:
x - y = 3
2x + y = 9
From the first equation, we express y using x.

y = x - 3

In the second equation, we write x-3 instead of y. And there we get a linear equation with one variable x.

2x + x - 3 = 9
3x = 9 + 3
3x = 12
x = 12/3
x = 4

Now that we found x, we can use it to find y.

y = x - 3
y = 4 - 3
y = 1

So, the solution of this system is (x,y) = (4,1).

Let's solve one more system using a different method:

Solve:

5x - 3y = 17
x + 3y = 11

5x - 3y + x + 3y = 17 - 11

Notice that we have -3y in the first equation and +3y in the second. If we add these 2, we get zero, which means we lose variable y. So, we add these 2 equations and we get a linear equation with one variable.

6x = 6
x = 1
Now that we have x, we use it to find y.

5 - 3y = 17
-3y = 17 - 5
-3y = 12
y = 12/(-3)
y = -4

Basic Algebra Solutions - Video Tutorials

https://test-preparation.ca/algebra-practice-questions/

1. 5x + 3 = 7x − 1. Find x

a. 1/3
b. ½
c. 1
d. 2

https://youtu.be/3v6_GBUSuE4

Solution

1. D
5x + 3 = 7x − 1
first we need to collect like terms
second is to undo addition and subtraction
third and the last is to undo multiplication and division
so again the first step is to put 5x and 7x in one side = 7x-5x
second put 3 and 1 in one side = 7x − 5x = 3 + 1
now we will solve the answer = 2x = 4
now we will divide by 2 to the both sides
2x/2 = 4/2
the answer now is x=2

2. $5x + 2(x + 7) = 14x - 7$. Find

a. 1
b. 2
c. 3
d. 4

https://youtu.be/xRccgQvkz_8

Solution

2. C
$5x + 2(x + 7) = 14x - 7$
$5x + 2x + 14 = 14x - 7$
$7x + 14 = 14x - 7$
$7x - 14x = -14 - 7$
$-7x = -21$
$x = 3$

3. $12t - 10 = 14t + 2$. Find t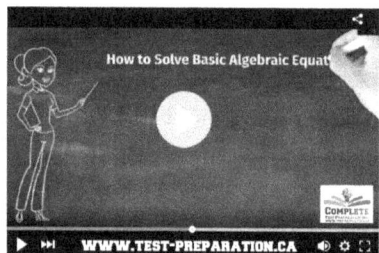

a. -6
b. -4
c. 4
d. 6

https://youtu.be/d9dsKx7U0Tg

3. A
$12t - 10 = 14t + 2$
$12t - 14t = 10 + 2$
$-2t = 12$
$-t = 6$
$t = -6$

4. $5(z + 1) = 3(z + 2) + 11$ Solve for Z

a. 2
b. 4
c. 6
d. 12

Solution

https://youtu.be/jWGAPCV1zRI

4. C
$5(z + 1) = 3(z + 2) + 11$. Z=?
$5z + 5 = 3z + 6 + 11$
$5z + 5 = 3z + 17$
$5z = 3Z + 17 - 5$
$5z - 3z = 12$
$2z = 12$
$z = 6$

Common Mistakes Answering Linear Equation Questions

Misinterpreting the question or setting up the equation Incorrectly

Failing to properly understand what the question is asking. Confusing linear equations with other types of algebra.

Misreading the problem and setting up the wrong equation.

Basic arithmetic errors: Addition, subtraction, multiplication, or division errors.
Basic Math Practice
https://test-preparation.ca/basic-math-practice-2/

Incorrectly applying the order of operations (PEMDAS). Order of Operation Practice
https://test-preparation.ca/order-of-operation/

Solving for the Wrong Variable:

Sign Errors: Incorrectly using positive and negative signs. Failing to change the sign when performing basic operations.

Fractions and Decimals: Incorrectly converting between fractions, decimals, and percentages.
Factions, decimals and percent practice

Incorrect Distribution: Mistakes when applying the distributive property. Forgetting to distribute a negative sign through parentheses.

Incorrectly Combining Like Terms: Failing to properly combine like terms.

Mistaking terms that are not alike for like terms.

Skipping Steps: Skipping steps in the process, leading to errors in the solution.

Not showing work, making it hard to identify where a mistake was made.

Checking Solutions: Forgetting to plug the solution back into the original equation to check its correctness.

Assuming an answer is correct without verification.

Mistakes Graphing: Errors plotting or connecting points.
Misinterpreting the slope and intercept in the line equation.

Time Management: Spending too much time on one problem and rushing through others. Not allocating enough time to review and check answers.
Time Management on a Test
https://test-preparation.ca/test-tactics-the-time-wise-approach/

Anxiety and Pressure: Feeling overwhelmed or stressed causing incorrect answers. Making careless mistakes due to test anxiety. Test Anxiety Tips
https://test-preparation.ca/how-to-overcome-test-anxiety/

Geometry

Cartesian Plane, Coordinate Grid and Plane

To locate points and draw lines and curves, we use the coordinate plane. It also called Cartesian coordinate plane. It is a two-dimensional surface with a coordinate grid in it, which helps us to count the units. For the counting of those units, we use x-axis (horizontal scale) and y-axis (vertical scale).

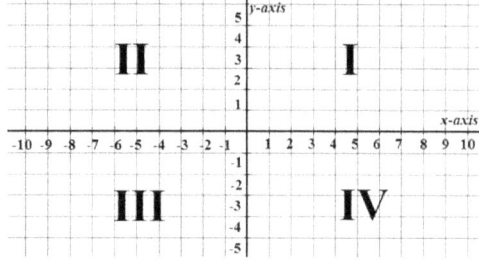

The whole system is called a coordinate system which is divided into 4 parts, called quadrants. The quadrant where all numbers are positive is the 1st quadrant (I), and if we go counterclockwise, we mark all 4 quadrants.

The location of a dot in the coordinate system is represented by coordinates. Coordinates are represented as a pair of numbers, where the 1st number is located on the x-axis and the 2nd number is located on the y-axis.

So, if a dot A has coordinates a and b, then we write:

A = (a,b)   or   A(a,b)

The point where x-axis and y-axis intersect is called an origin. The origin is the point from which we measure the distance along the x and y axes.

In the Cartesian coordinate system we can calculate the distance between 2 given points. If we have dots with coordinates:

A = (a,b)
B = (c,d)

Then the distance d between A and B can be calculated by the following formula:

$$d = \sqrt{(c-a)^2 + (d-b)^2}$$

Cartesian coordinate system is used for the drawing of 2-dimentional shapes, and is also commonly used for functions.

Example:

Draw the function   y = (1 - x)/2

To draw a linear function, we need at least 2 points. If we put that x = 0 then value for y would be:

$$y = \frac{1-x}{2} = \frac{1-0}{2} = \frac{1}{2}$$

We found the 1st point, let's name it A, with following coordinates:

A = (0,1/2)

To find the 2nd point, we can put that x=1. Here, the value for y would be:

$$y = \frac{1-x}{2} = \frac{1-1}{2} = \frac{0}{2} = 0$$

If we denote the 2nd point with B, then the coordinates for this point are:

B=(1,0)

Since we have 2 points necessary for the function, we find them in the coordinate system and we connect them with a line that represents the function,

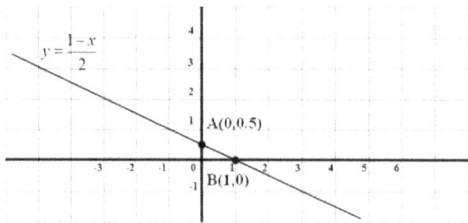

—— Geometry ——

Perimeter Area and Volume

Perimeter and Area (2-dimentional shapes)

Perimeter of a shape determines the length around that shape, while the area includes the space inside the shape.

Rectangle:

$P = 2a + 2b$
$A = ab$

Square

$P = 4a$
$A = a^2$

Parallelogram

$P = 2a + 2b$
$A = ah_a = bh_b$

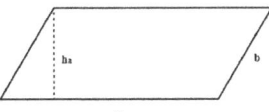

Rhombus

$P = 4a$
$A = ah = \dfrac{d_1 d_2}{2}$

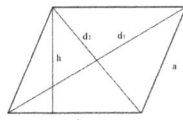

Triangle

$P = a+b+c$

$A = \dfrac{ah_a}{2} = \dfrac{bh_b}{2} = \dfrac{ch_c}{2}$

Equilateral Triangle

$P = 3a$

$A = \dfrac{a^2\sqrt{3}}{4}$

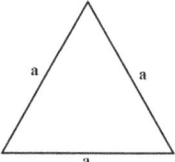

Trapezoid

$P = a+b+c+d$

$A = \dfrac{a+b}{2}h$

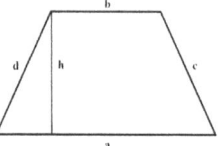

Circle

$P = 2r\pi$

$A = r^2\pi$

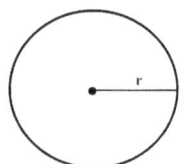

Area and Volume (3-dimentional shapes)

To calculate the area of a 3-dimentional shape, we calculate the areas of all sides and then we add them all.

To find the volume of a 3-dimentional shape, we multiply the area of the base (B) and the height (H) of the 3-dimentional shape.

$$V = BH$$

In case of a pyramid and a cone, the volume would be divided by 3.

$$V = BH/3$$

Here are some of the 3-dimentional shapes with formulas for their area and volume:

Cuboids

$A = 2(ab + bc + ac)$
$V = abc$

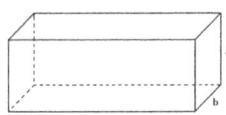

Cube

$A = 6a^2$
$V = a^3$

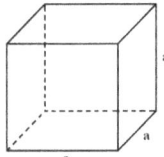

Pyramid

$A = ab + ah_a + bh_b$
$V = \dfrac{abH}{3}$

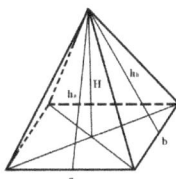

Cylinder

$A = 2r^2\pi + 2r\pi H$

$V = r^2\pi H$

Cone

$A = (r+s)r\pi$

$V = \dfrac{r^2\pi H}{3}$

https://youtu.be/ZWZ4NoCH7s8

https://youtu.be/UNPse0R9Qto

Adult Math Workbook

Pythagorean Geometry

If we have a right triangle ABC, where its sides (legs) are a and b and c is a hypotenuse (the side opposite the right angle), then we can establish a relationship between these sides using the following formula:

$$c^2 = a^2 + b^2$$

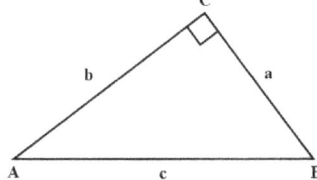

This formula is proven in the Pythagorean Theorem. There are many proofs of this theorem, but we'll look at just one geometrical proof:

If we draw squares on the right triangle's sides, then the area of the square upon the hypotenuse is equal to the sum of the areas of the squares that are upon other two sides of the triangle. Since the areas of these squares are a^2, b^2 and c^2, that is how we got the formula above.

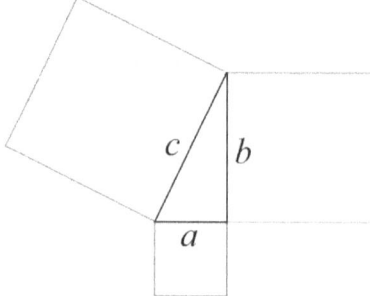

One of the famous right triangles is one with sides 3, 4

and 5. And we can see here that:

$3^2 + 4^2 = 5^2$
$9 + 16 = 25$
$25 = 25$

Example Problem:

The isosceles triangle ABC has a perimeter of 18 centimeters, and the difference between its base and legs is 3 centimeters. Find the height of this triangle.

We write the information we have about triangle ABC and we draw a picture for a better understanding of the rela-

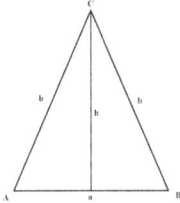

tion between its elements:

P=18 cm
a - b = 3 cm
h=?

We use the formula for the perimeter of the isosceles triangle, since that is what is given to us:
P = a + 2b = 18 cm

Notice that we have 2 equations with 2 variables, so we can solve it as a system of equations:

a + 2b = 18
a − b = 3 / a + 2b = 18
2a - 2b = 6 / a + 2b + 2a - 2b = 18 + 6

3a = 24
a = 24/3 = 8 cm

Now we go back to find b:
a - b = 3
8 - b = 3
b = 8 - 3
b = 5 cm

Using Pythagorean Theorem, we can find the height using a and b, because the height falls on the side a at the right angle. Notice that height cuts side a exactly in half, and that's why we use in the formula a/2. Here, b is our hypotenuse, so we have:

$b^2 = (a/2)^2 + h^2$
$h^2 = b^2 - (a/2)^2$
$h^2 = 5^2 - (8/2)^2$
$h^2 = 5^2 - (8/2)^2$
$h^2 = 25 - 4^2$
$h^2 = 26 - 16$
$h^2 = 9$
h = 3 cm.

Quadrilaterals

Quadrilaterals are 2-dimentional geometrical shapes that have 4 sides and 4 angles. There are many types of quadrilaterals, depending on the length of its sides, if they are parallel, and the size of its angles. All quadrilaterals have the following properties:

Sum of all interior angles is 360°

Sum of all exterior angles is 360°

Geometry

A quadrilateral is a parallelogram is it fulfills at least one of the following conditions:

- Angles on each side are supplementary
- Opposite angles are equal
- Opposite sides are equal
- Diagonals intersect each other exactly in half

Here are some of the quadrilaterals:

Square

All sides are equal
All angles are right angles

Rectangle

2 pairs of equal sides
All angles are right angles

Parallelogram

2 pairs of equal sides
Opposite angles are equal

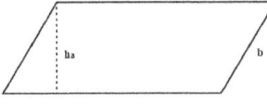

Rhombus

All sides are equal
Opposite angles are equal

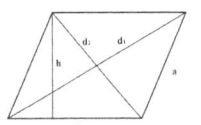

Trapezoid

One pair of parallel sides

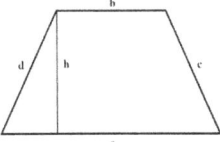

Example Problem
Find all angles of a parallelogram if one angle is greater than the other one by 40°.

First, we draw an image of a parallelogram:

We denote angles by α and β, Since this is a parallelogram, the opposite angles are equal.

We are given that one angle is greater than the other one by 40°, so we can write:

β = α + 40°

We solve this problem in two ways:
1) The sum of all internal angles of every quadrilateral is 360°. There are 2 α and 2 β. So we have:
2α + 2β = 360°
Now, instead of β we write α + 40:
2 α + 2 (α + 40°) =360°
2 α + 2 α + 80° = 360°
4 α = 360° - 80°
4 α = 280°
α = 280° / 4
α = 70°

Now we can find β from α:
β = α + 40°
β = 70° + 40°
β = 110°

2) One condition for parallelogram is "Angles on each side are supplementary" and we can use that to find these angles:

$\alpha + \beta = 180°$
$\alpha + \alpha + 40° = 180°$
$2\alpha = 180° - 40°$
$2\alpha = 140°$
$\alpha = 70°$

Now we find β:
$\beta = \alpha + 40°$
$\beta = 70° + 40°$
$\beta = 110°$

Common Mistakes in Geometry — What Students Often Get Wrong (and How to Fix Them)

Misreading or Mis-labeling Diagrams
Students often assume diagrams are drawn to scale or overlook labeling. Lines that look parallel may not be, angles may be misleading, and lengths may not appear proportionally accurate.

Tip: Always read the text, mark all given measurements, and recreate or label your own diagram instead of relying on visual intuition.

Ignoring or Omitting Units
Geometry often involves length, area, and volume—each with a distinct unit. A common slip is writing "cm" instead of "cm²" for area or forgetting units altogether.

Tip: Always pause to check that your answer includes appropriate units; this helps validate your reasoning.

Confusing Similarity and Congruence
Equating similar triangles with congruent ones is a common mistake — students expect proportional but identical sides, even when the two figures are scaled differently.

Remember: Similar = same shape (proportional sides); Congruent = same shape AND same size.

Applying Formulas Incorrectly
Students often confuse formulas — for example, mixing up perimeter with area, or using the wrong expressions for volume vs. surface area.

Tip: Identify the shape and ask: "What am I looking for—length, area, or volume?" Then select the formula that matches both shape and unit.

Misinterpreting Angle Relationships
Mistakes arise when handling angles formed by parallel lines, transversals, or in polygons. Students often forget angle relationships like alternate interior angles or the sum of interior angles of an n-sided polygon.

Tip: Quickly jot standard relationships: parallel line → corresponding/alternate interiors equal; polygon sum formula = $(n-2)\cdot 180°$; then check against the diagram.

Relying on Estimation Instead of Exact Reasoning
Estimation might help with eliminating multiple choice options, but in contexts like circle arcs or angled polygons, it's unreliable. Many test questions demand precision.

Tip: Use estimation only to check plausibility, not to choose an answer. Never skip the exact steps.

Geometry

Skipping Steps or Not Showing Work
In timed tests, students often jump to answers without jotting down intermediate steps.

Tip: Even if time is tight, write brief notes or formulas used. That helps prevent slip-ups and makes checking easier.

Not Using All the Given Information
Geometry problems are carefully designed—every given measure or relationship plays a role. Ignoring angles, lengths, or hints in the text leads to incomplete solutions.

Tip: Circle or list all given information right away. If your answer seems off, check whether you used all the given information.

Making Errors Under Pressure
Misplacing negative signs, dropping exponents, or miscopying diagram labels when anxious or rushed.

Tip: Underline key words, read questions twice, and re-confirm your final result before bubbling it in. Circle tough questions and revisit later with fresh eyes.

Adult Math Workbook

Answer Sheet

	A	B	C	D	E		A	B	C	D	E
1	○	○	○	○	○	21	○	○	○	○	○
2	○	○	○	○	○	22	○	○	○	○	○
3	○	○	○	○	○	23	○	○	○	○	○
4	○	○	○	○	○	24	○	○	○	○	○
5	○	○	○	○	○	25	○	○	○	○	○
6	○	○	○	○	○	26	○	○	○	○	○
7	○	○	○	○	○	27	○	○	○	○	○
8	○	○	○	○	○	28	○	○	○	○	○
9	○	○	○	○	○	29	○	○	○	○	○
10	○	○	○	○	○	30	○	○	○	○	○
11	○	○	○	○	○						
12	○	○	○	○	○						
13	○	○	○	○	○						
14	○	○	○	○	○						
15	○	○	○	○	○						
16	○	○	○	○	○						
17	○	○	○	○	○						
18	○	○	○	○	○						
19	○	○	○	○	○						
20	○	○	○	○	○						

Geometry Practice Questions

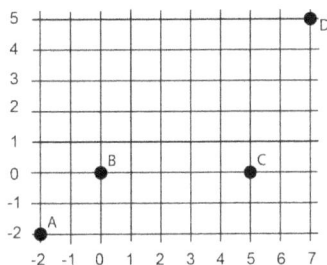

1. Which of the above points represents the origin?

   a. A
   b. B
   c. C
   d. D

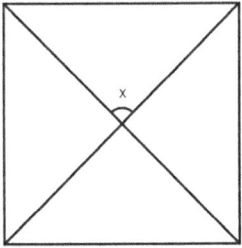

2. What is measurement of the indicated angle?

   a. 45°
   b. 90°
   c. 60°
   d. 30°

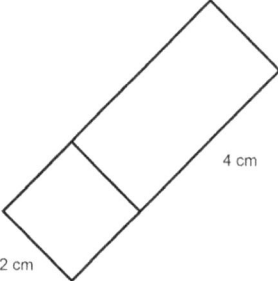

Note: figure not drawn to scale.

3. Assuming the figure with side 2 cm. is square, what is the perimeter of the above shape?

    a. 12 cm
    b. 16 cm
    c. 6 cm
    d. 20 cm

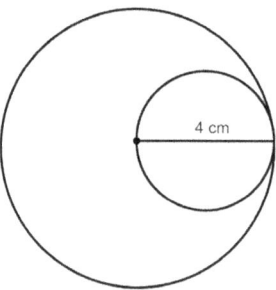

Note: Figure not drawn to scale

Geometry

4. Assuming the diameter of the small circle is the radius of the larger circle, what is (area of large circle) - (area of small circle) in the figure above?

    a. 8 π cm²
    b. 10 π cm²
    c. 12 π cm²
    d. 16 π cm²

Note: Figure not drawn to scale

5. Assuming the shapes around the center right triangle are square, what is the length of each side of the indicated square above?

    a. 10
    b. 15
    c. 20
    d. 5

6. Choose the expression the figure represents.

    a. X ≤ 1
    b. X < 1
    c. X > 1
    d. X ≥ 1

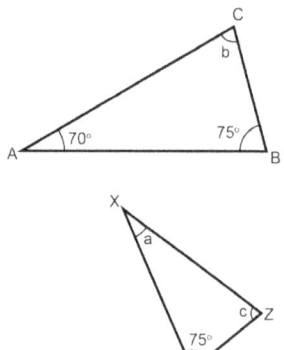

7. What are the respective values of a, b & c if both triangles are similar?

    a. 70°, 70°, 35°
    b. 70°, 35°, 70°
    c. 35°, 35°, 35°
    d. 70°, 75°, 35°

— Geometry —

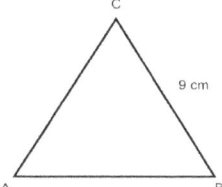

Note: figure not drawn to scale

8. What is the perimeter of the equilateral △ABC above?

    a. 18 cm
    b. 12 cm
    c. 27 cm
    d. 15 cm

Note: figure not drawn to scale

9. Assuming the 2 quadrangles are identical rectangles, what is perimeter of △ABC in the above shape?

    a. 25.5 cm
    b. 27 cm
    c. 30 cm
    d. 29 cm

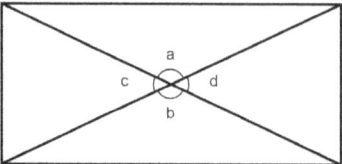

10. What is the sum of all the angles in the rectangle above?

    a. 180°
    b. 360°
    c. 90°
    d. 120°

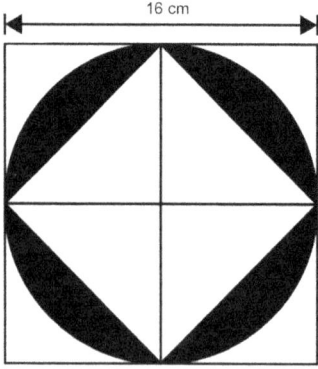

Note: figure not drawn to scale

Geometry

11. A tile factory makes custom tiles, shown above, from two types of stone. If a customer requires 200 tiles, how much black stone will be required?

   a. 256 m²
   b. 2560 m²
   c. 2.56 m²
   d. 25.6 m²

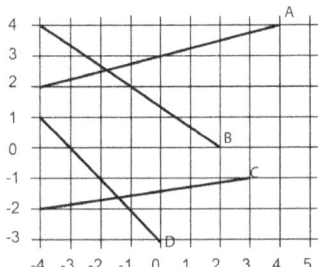

12. Which of the lines above represents the equation 2y − x = 4?

   a. A
   b. B
   c. C
   d. D

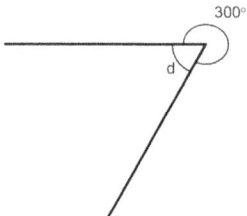

13. What is the measurement of the indicated angle?

    a. 45°
    b. 90°
    c. 60°
    d. 50°

5 cm

Note: figure not drawn to scale

14. What is the perimeter of the above shape, assuming the bottom portion is square?

    a. 22.85 cm
    b. 20 cm
    c. 15 cm
    d. 25.546 cm

Geometry

Note: Figure not drawn to scale

15. Assuming the quadrangle in the figure above is square, what is the length of the sides in the triangle above?

    a. 10
    b. 20
    c. 100
    d. 40

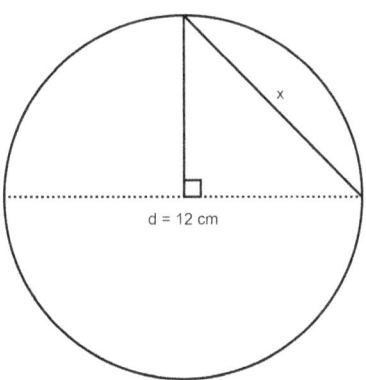

Note: Figure not drawn to scale

16. Calculate the length of side x.

   a. 6.46
   b. 8.48
   c. 3.6
   d. 6.4

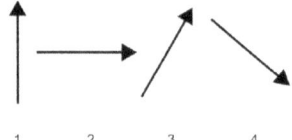

17. What is the correct order of respective slopes for the lines above?

   a. Positive, undefined, negative, positive
   b. Negative, zero, undefined, positive
   c. Undefined, zero, positive, negative
   d. Zero, positive undefined, negative

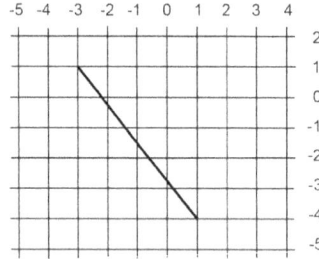

18. What is the slope of the line shown above?

   a. 5/4
   b. -4/5
   c. -5/4

d. -4/5

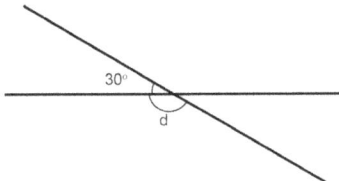

19. **What is the indicated angle above?**

    a. 150°
    b. 330°
    c. 60°
    d. 120°

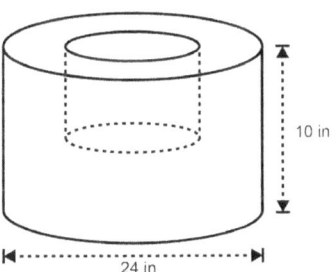

Note: figure not drawn to scale

20. What is the volume of the above solid made by a hollow cylinder that is half the size (in all dimensions) of the larger cylinder?

    a. 1440 π in³
    b. 1260 π in³
    c. 1040 π in³
    d. 960 π in³

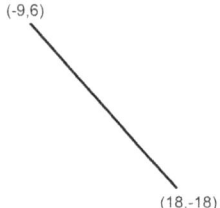

21. What is the slope of the line above?

    a. -8/9
    b. 9/8
    c. -9/8
    d. 8/9

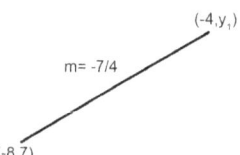

Geometry

22. With the data given above, what is the value of y_1?

    a. 0
    b. -7
    c. 7
    d. 8

Type A: 1300 ft²

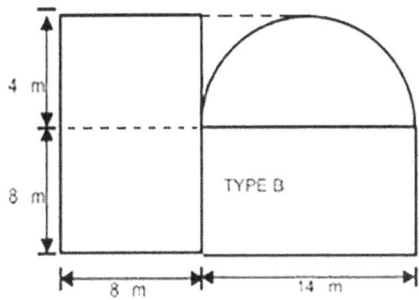

Note: Figure not drawn to scale

23. The price of houses in a certain subdivision is based on the total area. Susan is watching her budget and wants to choose the house with the lowest area. Which house type, A (1300 ft2) or B, should she choose if she would like the house with the lowest price? (1 m² = 10.76 ft² & π = 22/7)

a. Type B is smaller at 140 ft²
b. Type A is smaller
c. Type B is smaller at 855 ft²
d. Type B is larger

24. How much water can be stored in a cylindrical container 5 meters in diameter and 12 meters high?

Note: figure not drawn to scale

a. 235.65 m³
b. 223.65 m³
c. 240.65 m³
d. 252.65 m³

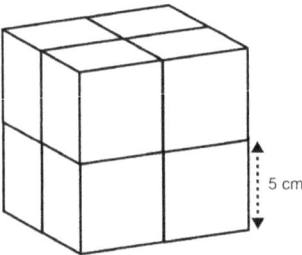

Note: figure not drawn to scale

Geometry

25. Assuming the figure above is composed of cubes, what is the volume?

a. 125 cm³
b. 875 cm³
c. 1000 cm³
d. 500 cm³

26. Choose the expression the figure represents.

a. X > 2
b. X ≥ 2
c. X < 2
d. X ≤ 2

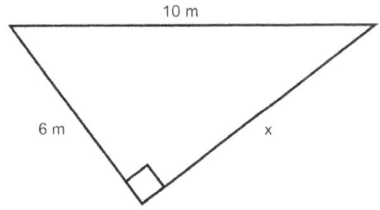

Note: figure not drawn to scale

27. What is the length of the missing side in the triangle above?

    a. 6
    b. 4
    c. 8
    d. 5

28. What is the value of the angle y?

    a. 25°
    b. 15°
    c. 30°
    d. 105°

29. What is the distance between the two points?

    a. ≈19
    b. 20
    c. ≈21
    d. ≈22

Geometry

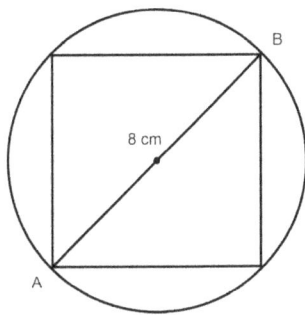

Note: figure not drawn to scale

30. What is area of the circle?

    a. 4 π cm²
    b. 12 π cm²
    c. 10 π cm²
    d. 16 π cm²

Answer Key

1. A

Point A represents the origin.

2. A

The diagonals of a square intersect at right angles, so each angle measures 90°. Half of that angle will be 45°

3. B

We see that there is a square with side 2 cm and a rectangle adjacent to it, with one side 2 cm (common side with the square) and the other side 4 cm. The perimeter of a shape is found by summing up all sides surrounding the shape, not adding the ones inside the shape. Three 2 cm sides from the square, and two 4 cm sides and one 2 cm side from the rectangle contribute the perimeter.

So, the perimeter of the shape is: 2 + 2 + 2 + 4 + 2 + 4 = 16 cm.

4. C

We are given a large circle and a small circle inside it; with the diameter equal to the radius of the large one. The diameter of the small circle is 4 cm. This means that its radius is 2 cm. Since the diameter of the small circle is the radius of the large circle, the radius of the large circle is 4 cm. The area of a circle is calculated by: πr^2 where r is the radius.

Area of the small circle: $\pi(2)^2 = 4\pi$

Area of the large circle: $\pi(4)^2 = 16\pi$

The difference area is found by:

Area of the large circle - Area of the small circle = $16\pi - 4\pi = 12\pi$

— Geometry —

5. B
We see that there are three squares forming a right triangle in the middle. Two of the squares have the areas 81 m² and 144 m². If we denote their sides a and b respectively:

a² = 81 and b² = 144. The length, which is asked, is the hypotenuse; a and b are the opposite and adjacent sides of the right angle. By using the Pythagorean Theorem, we can find the value of the asked side:

Pythagorean Theorem:

(Hypotenuse)² = (Opposite Side)² + (Adjacent Side)²

h² = a² + b²

a² = 81 and b² = 144 are given. So;

h² = 81 + 144

h² = 225

h = 15 m

6. B
The line is pointing towards numbers less than 1. The equation is therefore, X < 1.

7. D
Comparing respective angles - 70°, 75°, 35°

8. C
Equilateral triangle with 9 cm. sides
Perimeter = 9 + 9 + 9 = 27 cm.

9. D
Perimeter of triangle ABC is asked.
Perimeter of a triangle = sum of all three sides.

Here, Perimeter of △ABC = |AC| + |CB| + |AB|.

Since the triangle is located in the middle of two

adjacent and identical rectangles, we find the side lengths using these rectangles:

|AB| = 6 + 6 = 12 cm

|CB| = 8.5 cm

|AC| = |CB| = 8.5 cm

Perimeter = |AC| + |CB| + |AB| = 8.5 + 8.5 + 12 = 29 cm

10. B
a + b + c + d = ?
The sum of angles around a point is 360°
a + b + c + d = 360°

11. A
Black stone for 200 tiles = 200 x [Total tile area – Inner white area(4 triangles)]

= 200 x [(16^2) - (4 x 1/2 x 8 x 8)]

= 200 x (256 - 128) = 200 x 128 = 25600 cm²

Converting to meters – 1 cm. = 0.01 meters

= 25600/100 m²

= 256 m²

12. A
If a line represents an equation, all points on that line should satisfy the equation. Meaning that all (x, y) pairs present on the line should be able to verify that 2y - x is equal to 4. We can find out the correct line by trying a (x, y) point existing on each line. It is easier to choose points on the intersection of the grid lines:

Let us try the point (4, 4) on line A:

2 * 4 - 4 = 4

Geometry

8 - 4 = 4

4 = 4 ... this is a correct result, so the equation for line A is 2y - x = 4.

Let us try other points to check the other lines:

Point (-1, 2) on line B:

2 * 2 - (-1) = 4

4 + 1 = 4

5 = 4 ... this is a wrong result, so the equation for line B is not 2y - x = 4.

Point (3, -1) on line C:

2 * (-1) - 3 = 4

-2 - 3 = 4

-5 = 4 ... this is a wrong result, so the equation for line C is not 2y - x = 4.

Point (-2, -1) on line D:

2 * (-1) - (-2) = 4

-2 + 2 = 4

0 = 4 ... this is a wrong result, so the equation for line D is not 2y - x = 4.

13. C
The sum of angles around a point is 360°

d + 300 = 360°

d = 60°

14. A
Find the perimeter of a shape made by merging a square and a semi circle. Perimeter = 3 sides of the square + 1/2 circumference of the circle.

= (3 x 5) + 1/2 (5 π)

= 15 + 2.5 π

= 15 + 7.853975

Perimeter = 22.85 cm

15. A
If we call one side of the square "a," the area of the square will be a^2.

We know that a^2 = 200 cm².

On the other hand; there is an isosceles right triangle. Using the **Pythagorean Theorem:**

(Hypotenuse)² = (Adjacent Side)² + (Opposite Side)²
Where the hypotenuse is equal to one side of the square. So,

$a^2 = x^2 + x^2$

$200 = 2x^2$

$200/2 = 2x^2/2$

$100 = x^2$

$x = \sqrt{100}$

x = 10 cm

16. B
In the question, we have a right triangle formed inside the circle. We are asked to find the length of the hypotenuse of this triangle. We can find the other two sides of the triangle by using circle properties:

The diameter of the circle is equal to 12 cm. The legs of the right triangle are the radii of the circle; so they are 6 cm long.

Geometry

Using the Pythagorean Theorem:

$(\text{Hypotenuse})^2 = (\text{Adjacent Side})^2 + (\text{Opposite Side})^2$

$x^2 = r^2 + r^2$

$x^2 = 6^2 + 6^2$

$x^2 = 72$

$x = \sqrt{72}$

$x = 8.48$

17. C
Undefined, zero, positive, negative.

18. C
Slope (m) = $\dfrac{\text{change in y}}{\text{change in x}}$

$(x_1, y_1) = (-3, 1)$ & $(x_2, y_2) = (1, -4)$

Slope = $[-4 - 1]/[1-(-3)] = -5/4$

19. A
The angles opposite both angles 30° and angle d are respectively equal to vertical angles.

$2(30° + d) = 360°$

$2d = 360° - 60°$

$2d = 300°$

$d = 150°$

20. B
Total Volume = Volume of large cylinder - Volume of small cylinder

Volume of a cylinder = area of base • height = $\pi r^2 \cdot h$

Total Volume = $(\pi * 12^2 * 10) - (\pi * 6^2 * 5) = 1440\pi - 180\pi$

= 1260π in³

21. A
If we know the coordinates of two points on a line, we can find the slope (m) with the below formula:

$m = (y_2 - y_1)/(x_2 - x_1)$ where (x_1, y_1) represent the coordinates of one point and (x_2, y_2) the other.

In this question:

$(-9, 6) : x_1 = -9, y_1 = 6$

$(18, -18) : x_2 = 18, y_2 = -18$

Inserting these values into the formula:

$m = (-18 - 6)/(18 - (-9)) = (-24)/(27)$ ... Simplifying by 3:

$m = -8/9$

22. A
If we know the coordinates of two points on a line, we can find the slope (m) with the below formula:

$m = (y_2 - y_1)/(x_2 - x_1)$ where (x_1, y_1) represent the coordinates of one point and (x_2, y_2) the other.

In this question:

$(-4, y_1) : x_1 = -4, y_1 =$ we will find

$(-8, 7) : x_2 = -8, y_2 = 7$

$m = -7/4$

Inserting these values into the formula:

$-7/4 = (7 - y_1)/(-8 - (-4))$

$-7/4 = (7 - y_1)/(-8 + 4)$

$7/(-4) = (7 - y_1)/(-4)$ ... Simplifying the denominators of both sides by -4:

$7 = 7 - y_1$

$0 = -y_1$

$y_1 = 0$

23. D
Area of Type B consists of two rectangles and a half circle. We can find these three areas and sum them up to find the total area:

Area of the left rectangle: (4 + 8) * 8 = 96 m²

Area of the right rectangle: 14 * 8 = 112 m²

The diameter of the circle is equal to 14 m. So, the radius is 14/2 = 7:

Area of the half circle = (1/2) * πr² = (1/2) * (22/7) * (7)² = (1 * 22 * 49) / (2 * 7) = 77 m²

Area of Type B = 96 + 112 + 77 = 285 m²

Converting this area to ft²: 285 m² = 285•10.76 ft² = 3066.6 ft²

Type B is (3066.6 - 1300 = 1766.6 ft²) 1766.6 ft² larger than type A.

24. A
The formula of the volume of cylinder is the base area multiplied by the height. As the formula:

Volume of a cylinder = πr²h. Where π is 3.142, r is radius of the cross sectional area, and h is the height.

We know that the diameter is 5 meters, so the radius is 5/2 = 2.5 meters.

The volume is: V = 3.142 * 2.5² * 12 = 235.65 m³.

25. C
The large cube is made up of 8 smaller cubes with 5 cm sides. The volume of a cube is found by the third power of the length of one side.

Volume of the large cube = Volume of the small cube * 8

= $(5^3) * 8 = 125 * 8$

= 1000 cm³

There is another solution for this question. Find the side length of the large cube. There are two cubes rows with 5 cm length for each. So, one side of the large cube is 10 cm.

The volume of this large cube is equal to 10^3 = 1000 cm³

26. A
The line is pointing towards numbers greater than 2. The equation is therefore, X > 2.

27. C
Pythagorean Theorem:
(Hypotenuse)² = (Perpendicular)² + (Base)²

$h^2 = a^2 + b^2$

Given: a = 6, h = 10

$h^2 = a^2 + b^2$

$b^2 = h^2 - a^2$

$b^2 = 10^2 + 6^2$

$b^2 = 100 - 36$

$b^2 = 64$

b = 8

28. D

Two parallel lines intersected by a third line with angles of 75°

x = 75° (corresponding angles)
x + y = 180° (supplementary angles)

y = 180° - 75°

y = 105°

29. D

The distance between two points is found by
= $[(x_2 - x_1)^2+(y_2 - y_1)^2]^{1/2}$

In this question:

(18, 12) : x_1 = 18, y_1 = 12

(9, -6) : x_2 = 9, y_2 = -6

Distance= $[(9 - 18)^2 + (-6 - 12)^2]^{1/2}$

= $[(-9)^2 + (-18)^2]^{1/2}$

= $(9^2 + 2^2 \cdot 9^2)^{1/2}$

= $(9^2(1 + 5))^{1/2}$ ... We can take 9 out of the square root:

= $9 * 5^{1/2}$

= $9\sqrt{5}$

= 9 * 2.45

= 22.05

The distance is approximately 22 units.

30. D

We have a circle given with diameter 8 cm and a square located within the circle. We are asked to find the area of the circle for which we only need to know the length of the radius that is the half of the diameter.

Area of circle = πr^2 ... r = 8/2 = 4 cm

Area of circle = $\pi * 4^2$

= 16π cm^2 ... As we notice, the inner square has no role in this question.

Basic Math Multiple Choice

Math is the one section where you need to make sure that you understand the processes before you ever tackle it. That's because the time allowed on the math portion is typically so short that there's not much room for error. You have to be fast and accurate. It's imperative that before the test day arrives, you've learned all the main formulas that will be used, and then to create your own problems (and solve them).

On the actual test day, use the "Plug-Check-Check" strategy. Here's how it goes.

Read the problem, but not the answers. You'll want to work the problem first and come up with your own answers. If you did the work right, you should find your answer among the options given.

If you need help with the problem, plug actual numbers into the variables given. You'll find it easier to work with numbers than it is to work with letters. For instance, if the question asks, "If Y-4 is 2 more than Z, then Y+5 is how much more than Z?" try selecting a value for Y. Let's take 6. Your question now becomes, "If 6-4 is 2 more than Z, then 6 plus 5 is how much more than Z?" Now your answer should be easier to work with.

Check the answer choices to see if your answer matches one of those. If so, select it.

If no answer matches the one you got, re-check your math, but this time, use a different method. In math, it's common for there to be more than one way to solve a problem. As a simple example, if you multiplied 12 X 13 and did not get an answer that matches one of the answer choices, you might try adding 13 together 12 different times and see if you get a good answer.

Math Multiple Choice Strategy

The two strategies for working with basic math multiple choice are Estimation and Elimination.

Math Strategy 1 - Estimation.

Just like it sounds, try to estimate an approximate answer first. Then look at the choices.

Math Strategy 2 - Elimination.

For every question, no matter what type, eliminating obviously incorrect answers narrows the possible choices. Elimination is probably the most powerful strategy for answering multiple choice.

Here are a few basic math examples.

Solve 2/3 + 5/12

    a. 9/17

    b. 3/11

    c. 7/12

    d. 1 1/12

Basic Math Multiple Choice

First estimate the answer. 2/3 is more than half and 5/12 is about half, so the answer is going to be very close to 1.

Next, Eliminate. Choice A is about 1/2 and can be eliminated, choice B is very small, less than 1/2 and can be eliminated. Choice C is close to 1/2 and can be eliminated. Leaving only choice D, which is just over 1.

Work through the solution, a common denominator is needed, a number which both 3 and 12 will divide into. 2/3 = 8/12. So, 8+5/12 = 13/12 = 1 1/12

Choice D is correct.

Solve 4/5 – 2/3

    a. 2/2
    b. 2/13
    c. 1
    d. 2/15

You can eliminate choice A, because it is 1, and since both numbers are close to one, the difference is going to be very small. You can eliminate choice C for the same reason.

Next, look at the denominators. Since 5 and 3 don't go into 13, you can eliminate choice B as well.

That leaves choice D.

Checking the answer, the common denominator will be 15. So 12-10/15 = 2/15. Choice D is correct.

Fractions Shortcut - Canceling Out

In any operation with fractions, if the numerator of one fraction has a common multiple with the denominator of the other, you can cancel out. This saves time and simplifies the problem quickly, making it easier to manage.

Solve 2/15 ÷ 4/5

a. 6/65

b. 6/75

c. 5/12

d. 1/6

To divide fractions, we multiply the first fraction with the inverse of the second fraction. Therefore we have 2/15 x 5/4. The numerator of the first fraction, 2, shares a multiple with the denominator of the second fraction, 4, which is 2. These cancel out, which gives, 1/3 x 1/2 = 1/6

Canceling Out solved the questions very quickly, but we can still use multiple choice strategies to answer.

Choice B can be eliminated because 75 is too large a denominator. Choice C can be eliminated because 5 and 15 don't go into 12.

Choice D is correct.

Decimal Multiple Choice Strategy and Shortcuts

Multiplying decimals gives a very quick way to estimate and eliminate choices. Anytime that you multiply decimals, it is going to give an answer with the same number of decimal places as the combined operands.

So for example,

2.38 X 1.2 will produce a number with three places of decimal, which is 2.856.
Here are a few examples with step-by-step explanation:

Solve 2.06 x 1.2

    a. 24.82

    b. 2.482

    c. 24.72

    d. 2.472

This is a simple question, but even before you start calculating, you can eliminate several choices. When multiplying decimals, there will always be as many numbers behind the decimal place in the answer as the sum of the ones in the initial problem, so choices A and C can be eliminated.

The correct answer is D: 2.06 x 1.2 = 2.472

Solve 20.0 ÷ 2.5

a. 12.05
b. 9.25
c. 8.3
d. 8

First estimate the answer to be around 10, and eliminate choi And since it'd also be an even number, you can eliminate choic and C, leaving only choice D.

The correct Answer is D: 20.0 ÷ 2.5 = 8

How to Study for a Math Test

EVERY SUBJECT HAS ITS OWN PARTICULAR STUDY METHOD. Math is mostly numerical, rather than verbal, and requires logical thinking; it has its own way to be studied. Before touching on significant points of studying a math test, lets look at some of the fundamentals of "learning."

Learning is not an instant experience; it is a procedure. Learning is a process not an event. Rome wasn't built in a day, and learning anything (or everything) isn't going to happen in a day either. You cannot expect to learn everything in one day, at night, before the test. It is important and necessary to learn day-by-day. Good time management plays a considerable role in learning. When you manage your time, and begin test preparation well in advance, you will notice the subjects are easier than you thought, or feared, and you will take the test without the stress of a sleepless body and an anxious mind.

Memorizing is a temporary step of learning if information is not comprehended and applied afterwards. Memorize just the basics and understand the meaning; then apply, analyze, synthesize and evaluate.

These are the hierarchical layout of cognitive learning: Of course, there are some basic properties that you need to memorize in the beginning, since you cannot prove the facts every time you solve a math test. For example; the inner angles of a triangle sum up to 180°. If you do not know this, you may not be able to solve triangle problems. And, more important, if you do not practice, you will certainly forget it. Practice helps information take root in your brain. Applying the same property to various types of questions extends the roots.

For example, if you see a triangle, you can analyze the question by the property. In a question, if you see a hexagon, you can split it into triangles and use the property, called synthesizing followed by evaluation. Following these steps, the property is completely learned and has its place in your long term memory.

A useful method in providing consistent learning is using similarities between the information and events, images, shapes, ... etc. For example, assume that you have difficulty remembering the formula
$x^a/y^b = x^a y^{-b} = 1/(x^{-a} y^b)$.

You can associate this to an elevator: The exponents changing location (numerator/denominator) need to change exponent sign, similarly, people going up need to push the up button and if they decide to go down, they need to push the down button; so they need to change the button. Also; writing the formula in large letters and sticking it on a surface that is frequently visible helps memorizing it by using visual intelligence. The more senses (visual, musical, auditory, logical, ...) the material is associated with, the more permanent it is.

Attend to all classes. Knowledge is not replaceable by others, and every brain is unique. You cannot learn math

How to Study for a Math Test

from your classmate's notes; take your own notes in your own understanding of the material. What you understand, or don't understand, and how you understand it is different to everyone else. Highlight the important points in your own way. Remember math and all other courses are mostly learned at school - practice comes afterwards at home.

Find your own way of learning. Every person learns and studies differently. Some take notes, some do not like writing; listening is the major way of embracing information for them, and some watch. It is important to detect the way that is more useful for you. Coloring important points also helps. Due to selective perception; we see the attractive words, signs before the rest. While studying math, make a list; first, determine the subjects you feel inadequate on and focus on them initially.

Never gloss over something that you do not fully understand. Information is built on previous learning in a hierarchical order. If you have questions about a mathematical property, and don't understand is completely, you cannot solve problems using that property. This true for any subject, but especially for math. You need to have a strong background to succeed in math. You need to know your basic math inside out to do algebra. And you need to know your algebra inside out to do calculus. If you do not know exponentials, you cannot solve logarithms. If you don't understand something, get help from your teachers, reread course materials, follow the examples in the textbook from start to finish, discuss with friends or hire a private tutor. Never skip over something that you don't understand – it will come back to haunt you!

Practice makes perfect! Yes – it really does! Working through math problems in your own way is essential. Looking over examples is a good first step – but only that.

The example solutions shown in the textbook show you have to solve a problem, next you have to do it yourself. Completing assignments and practice questions are critical. You see different types of examples and acquire different outlooks when facing math problems. Math is fun because usually there are many ways to reach the solution. Find alternative ways to solve a problem to anchors your learning deeper. Find similar and different problems, discuss with friends, ask each other questions. Observing other people's way of thinking, and solving problems will help both of you improves.

Succeeding in math is a mental action. However, do not disregard physical and psychological effects. Always think positive and never give up; no success is gained without effort. Of course, you will waste time solving problems you find easy, and you will struggle with difficult problems. In the end, you will be one step further ahead. And then after more time spent practicing, you will be another step ahead.

Reward yourself after intense studies. This will keep your motivation high. The reward may be a chocolate bar, playing a game for 20 minutes, or taking a walk in the park. It is very essential to have a good night's sleep before taking a math test. Eating habits directly effect success. Keep away from fast food as much as you can, eat a light meal before a test. Finally, keep your inner motivation very high. Believe in yourself; you will certainly get the good result from planned, efficient studies.

How to Prepare for a Test

MOST STUDENTS HIDE THEIR HEADS AND PROCRASTINATE WHEN FACED WITH PREPARING FOR AN EXAM, HOPING THAT SOMEHOW THEY WILL BE SPARED THE AGONY, ESPECIALLY IF IT IS A BIG ONE THAT THEIR FUTURES RELY ON. Avoiding a test is what many students do best and unfortunately, they suffer the consequences because of their lack of preparation.

Test preparation requires strategy and dedication. It is the perfect training ground for a professional life. Besides having several reliable strategies, successful students also has a clear goal and know how to accomplish it. These tried and true concepts have worked well and will make your test preparation easier.

The Study Approach

Take responsibility for your own test preparation.

It is a common - but big - mistake to link your studying to someone else's. Study partners are great, but only if they are reliable. It is your job to be prepared for the test, even if a study partner fails you. Do not allow others to distract you from your goals.

Prioritize the time available to study

When do you learn best, early in the day or at night? Does your mind absorb and retain information most efficiently in small blocks of time, or do you require long stretches to get the most done? It is important to figure out the best blocks of time available to you when you can be the most productive. Try to consolidate activities to allow for longer periods of study time.

Find a quiet place where you will not be disturbed

 Do not try to squeeze in quality study time in any old location. Find a peaceful place with a minimum of distractions, such as the library, a park or even the laundry room. Good lighting is essential and you need to have comfortable seating and a desk surface large enough to hold your materials. It is probably not a great idea to study in your bedroom. You might be distracted by clothes on the floor, a book you have been planning to read, the telephone or something else. Besides, in the middle of studying, that bed will start to look very comfortable. Whatever you do, avoid using the bed as a place to study since you might fall asleep to avoiding studying!

The exception is flashcards. By far the most productive study time is sitting down and studying and studying only. However, with flashcards you can carry them with you and make use of odd moments, like standing in line or waiting for the bus. This isn't as productive, but it really helps and is definitely worth doing.

Determine what you need to study

Gather together your books, your notes, your laptop and any other materials needed to focus on your study for this exam. Ensure you have everything you need so you don't waste time. Remember paper, pencils and erasers, sticky notes, bottled water and a snack. Keep your phone with you if you need it to find essential information, but keep it turned off so others can't distract you.

Have a positive attitude

It is essential that you approach your studies for the test with an attitude that says you will pass it. And pass it with flying colors! This is one of the most important keys to successful studying. Believing that you are capable helps you to become capable.

The Strategy of Studying

Review class notes

Stay on top of class notes and assignments by reviewing them frequently and regularly. Re-writing notes can be a terrific study trick, as it helps lock in information. Pay special attention to any comments that have been made by the teacher. If a study guide has been made available as part of the class materials, use it! It will be a valuable tool to use for studying.

Estimate how much time you will need

If you are concerned about the amount of time you have available it is a good idea to set up a schedule so that you do not get bogged down on one section and end without enough time left to study other things. Remember to schedule break time, and use that time for a little exercise or other stress reducing techniques.

Test yourself to determine your weaknesses

Look online for additional assessment and evaluation tools available for a particular subject. Visit our website https://www.test-preparation.ca for test tips and more practice questions. Once you have determined areas of concern, you will be able to focus on studying the information they contain and just brush up on the other areas of the exam.

Mental Prep – How to Psych Yourself Up for a Test

Since tests are often a big factor in your final grade or acceptance into a program, it is understandable that taking tests creates anxiety for many students. Even students who know they have learned the required material find their minds going blank as they stare at the test booklet. One easy way to overcome that anxiety is to prepare mentally for the test. Here are a few simple techniques.

Do not procrastinate

Study the material for the test when it becomes available, and continue to review the material until the test day. By waiting until the last minute and trying to cram for the test the night before, you actually increase anxiety. This leads to an increase in negative self-talk. Telling yourself "I can't learn this. I am going to fail" is a pretty sure indication that you are right. At best, your performance on the test will not be as strong if you have procrastinated instead of studying.

Positive self-talk.

Positive self-talk drowns out negative self-talk and to increases your confidence level. Whenever you begin feeling overwhelmed or anxious about the test, remind yourself that you have studied enough, you know the material and that you will pass the test. Both negative and positive self-talk are really just your fantasy, so why not choose to be a winner?

Do not compare yourself to others.

Do not compare yourself to other students. Instead, focus on your strengths and weaknesses and prepare accordingly. Regardless of how others perform, your performance is the only one that matters to your grade. Comparing yourself to others increases your anxiety and negative self-talk before the test.

Visualize.

Make a mental image of yourself taking the test. You know the answers and feel relaxed. Visualize doing well on the test and having no problems with the material. Visualizations can increase your confidence and decrease the anxiety you might otherwise feel before the test. Instead of thinking of this as a test, see it as an opportunity to demonstrate what you have learned!

Avoid negativity.

Worry is contagious and viral - once it gets started it builds on itself. Cut it off before it gets to be a problem. Even if you are relaxed and confident, being around anxious, worried classmates might cause you to start feeling anxious. Before the test, tune out the fears of classmates. Feeling anxious and worried before an exam is normal, and every student experiences those feelings at some point. But you cannot allow these feelings to interfere with your ability to perform well. Practicing mental preparation techniques and remembering that the test is not the only measure of your academic performance will ease your anxiety and ensure that you perform at your best.

How to Take a Test

EVERYONE KNOWS THAT TAKING AN EXAM IS STRESSFUL, BUT IT DOES NOT HAVE TO BE THAT BAD! There are a few simple things that you can do to increase your score on any type of test. Take a look at these tips and consider how you can incorporate them into your study time.

OK - so you are in the test room - Here is what to do!

Reading the Instructions

This is the most basic point, but one that, surprisingly, many students ignore and it costs big time! Since reading the instructions is one of the most common, and 100% preventable mistakes, we have a whole section just on reading instructions.

Pay close attention to the sample questions. Almost all standardized tests offer sample questions, paired with their correct solutions. Go through these to make sure that you understand what they mean and how they arrived at the correct answer. Do not be afraid to ask the test supervisor for help with a sample that confuses you, or instructions that you are unsure of.

Tips for Reading the Question

We could write pages and pages of tips just on reading the test questions. Here are a few that will help you the most.

How to Take a Test

- **Think first.** Before you look at the answer, read and think about the question. It is best to try to come up with the correct answer before you look at the options. This way, when the test-writer tries to trick you with a close answer, you will not fall for it.

- **Make it true or false.** If a question confuses you, then look at each answer option and think of it as a "true" "false" question. Select the one that seems most likely to be "true."

- **Mark the Question.** For some reason, a lot of test-takers are afraid to mark up their test booklet. Unless you are specifically told not to mark in the booklet, you should feel free to use it to your advantage.

- **Circle Key Words.** As you are reading the question, underline or circle key words. This helps you to focus on the most critical information needed to solve the problem. For example, if the question said, "Which of these is not a synonym for huge?" You might circle "not," "synonym" and "huge." That clears away the clutter and lets you focus on what is important.

- **Always underline these words:** all, none, always, never, most, best, true, false and except.

- **Eliminate.** Elimination is the best strategy for multiple choice answers *and* questions. If you are confused by lengthy questions, cross out anything that you think is irrelevant, obviously wrong, or information that you think is offered to distract you.

- **Do not try to read between the lines.** Usually, questions are written to be straightforward, with no deep, underlying meaning. Generally, the simple answer really is the correct answer. Do not over-analyze!

How to Take a Test - The Basics

Some sections of the test are designed to assess your ability to quickly grab the necessary information; this type of exam makes speed a priority. Others are more concerned with your depth of knowledge, and how accurate it is. When you start a new section of the test, look it over to determine whether the test is for speed or accuracy. If the test is for speed (a lot of questions and a short time), your strategy is clear; answer as many questions as quickly as possible.

Most tests do NOT penalize for wrong answers, so if all else fails, guess and make sure you answer every question.

Every little bit helps

Most tests do NOT allow personal calculators. Make sure you know how to use it!

You cannot bring any other materials into the test room. Scratch paper and a pencil are provided. Use them!

Make time your friend

Budget your time from the beginning until you are finished, and stick to it! The amount of time you are permitted for each portion of the test will almost certainly be included in the instructions.

How to Take a Test

Easy does it

One smart way to tackle a test is to locate the easy questions and answer those first. This is a time-tested strategy that never fails, because it saves you a lot of unnecessary anxiety. First, read the question and decide if you can answer it in less than a minute. If so, complete the question and go to the next one. If not, skip it for now and continue to the next question. By the time you have completed the first pass through this section of the exam, you will have answered a good number of questions. Not only does it boost your confidence, relieve anxiety and kick your memory up a notch, you will know exactly how many questions remain and can allot the rest of your time accordingly. Think of doing the easy questions first as a warm-up!

Do not watch your watch

At best, taking an important exam is an uncomfortable situation. If you are like most people, you might be tempted to subconsciously distract yourself from the task at hand. One of the most common ways to do so is by becoming obsessed with your watch or the wall clock. Do not watch your watch! Take it off and place it on the top corner of your desk, far enough away that you will not be tempted to look at it every two minutes. Better still, turn the watch face away from you. That way, every time you try to sneak a peek, you will be reminded to refocus your attention to the task at hand. Give yourself permission to check your watch or the wall clock after you complete each section. Focus on answering the questions, not on how many minutes have elapsed since you last looked at it.

Divide and conquer

What should you do when you come across a question that is so complicated you may not even be certain what is being asked? As we have suggested, the first time through, skip the question. At some point, you will need to return to it and get it under control. The best way to handle questions that leave you feeling so anxious you can hardly think is by breaking them into manageable pieces. Solving smaller bits is always easier. For complicated questions, divide them into bite-sized pieces and solve these smaller sets separately. Once you understand what the reduced sections are really saying, it will be much easier to put them together and get a handle on the bigger question. This may not work with every question - see below for how to deal with questions you cannot break down.

Reason your way through the toughest questions

If you find that a question is so dense you can't figure out how to break it into smaller pieces, there are a few strategies that might help. First, read the question again and look for hints. Can you re-word the question in one or more different ways? This may give you clues. Look for words that can function as either verbs or nouns, and try to figure out what the questions is asking from the sentence structure. Remember that many nouns in English have several different meanings. While some of those meanings might be related, sometimes they are completely distinct. If reading the sentence one way does not make sense, consider a different definition or meaning for a key word.

The truth is, it is not always necessary to understand a question to arrive at a correct answer! The most successful strategy for multiple choice is Elimination. Frequently, at least one answer is clearly wrong and can be crossed off the list of possible correct answers. Next, look at the remaining answers and eliminate any that are only partially true. You may still have to flat-out guess from time to time, but using the process of elimination will help you make your way to the correct answer more often than not - even when you don't know what the question means!

Do not leave early

Use all the time allotted to you, even if you can't wait to get out of the testing room. Instead, once you have finished, spend the remaining time reviewing your answers. Go back to those questions that were most difficult for you and review your response. Another good way to use this time is to return to multiple-choice questions in which you filled in a bubble. Do a spot check, reviewing every fifth or sixth question to make sure your answer coincides with the bubble you filled in. This is a great way to catch yourself if you made a mistake, skipped a bubble and therefore put all your answers in the wrong bubbles!

Become a super sleuth and look for careless errors. Look for questions that have double negatives or other odd phrasing; they might be an attempt to throw you off. Careless errors on your part might be the result of skimming a question and missing a key word. Words such as "always", "never", "sometimes" , "rarely" and the like can give a strong indication of the answer the question is really seeking. Don't throw away points by being careless!

Just as you budgeted time at the beginning of the test to allow for easy and more difficult questions, be sure to budget sufficient time to review your answers. On essay questions and math questions where you are required to show your work, check your writing to make sure it is legible.

Math questions can be especially tricky. The best way to double check math questions is by figuring the answer using a different method, if possible.

Here is another terrific tip. It is likely that no matter how hard you try, you will have a handful of questions you just are not sure of. Keep them in mind as you read through the rest of the test. If you can't answer a question, looking back over the test to find a different question that addresses the same topic might give you clues.

We know that taking the test has been stressful and you can hardly wait to escape. Just keep in mind that leaving before you double-check as much as possible can be a quick trip to disaster. Taking a few extra minutes can make the difference between getting a bad grade and a great one. Besides, there will be lots of time to relax and celebrate after the test is turned in.

In the Test Room – What you MUST do!

If you are like the rest of the world, there is almost nothing you would rather avoid than taking a test. Unfortunately, that is not an option if you want to pass. Rather than suffer, consider a few attitude adjustments that might turn the experience from a horrible one to...well, an interesting one! Take a look at these tips. Simply changing how you perceive the experience can change the experience itself.

You have to take the test - you can't change that. What you can change, and the only thing that you can change, is your attitude -so get a grip - you can do it!

Get in the mood

After weeks of studying, the big day has finally arrived. The worst thing you can do to yourself is arrive at the test site feeling frustrated, worried, and anxious. Keep a check on your emotional state. If your emotions are shaky before a test it can determine how well you do on the test. It is extremely important that you pump yourself up, believe in yourself, and use that confidence to get in the mood!

Don't fight reality

Students often resent tests, and with good reason. After all, many people do not test well, and they know the grade they end with does not accurately reflect their true knowledge. It is easy to feel resentful because tests classify students and create categories that just don't seem fair. Face it: Students who are great at rote memorization and not that good at actually analyzing material often score higher than those who might be more creative thinkers and balk at simply memorizing cold, hard facts. It may not be fair, but there it is anyway. Conformity is an asset on tests, and creativity is often a liability. There is no point in wasting time or energy being upset about this reality. Your first step is to accept the reality and get used to it. You will get higher marks when you realize tests do count and that you must give them your best effort. Think about your future and the career that is easier to achieve if you have consistently earned high grades. Avoid negative energy and focus on anything that lifts your enthusiasm and increases your motivation.

Get there early enough to relax

If you are wound up, tense, scared, anxious, or feeling rushed, it will cost you. Get to the exam room early and relax before you go in. This way, when the exam starts, you are comfortable and ready to apply yourself. Of course, you do not want to arrive so early that you are the only one there. That will not help you relax; it will only give you too much time to sit there, worry and get wound up all over again.

If you can, visit the room where you will be taking your exam a few days ahead of time. Having a visual image of the room can be surprisingly calming, because it takes away one of the big 'unknowns'. Not only that, but once you have visited, you know how to get there and will not be worried about getting lost. Furthermore, driving to the test site once lets you know how much time you need to allow for the trip. That means three potential stressors have been eliminated all at once.

Get it down on paper

One advantages of arriving early is that it allows you time to recreate notes. If you spend a lot of time worrying about whether you will be able to remember information like names, dates, places, and mathematical formulas, there is a solution for that. Unless the exam you are taking allows you to use your books and notes, (and very few do) you will have to rely on memory. Arriving early gives to time to tap into your memory and jot down key pieces of information you know will be asked. Just make certain you are allowed to make notes once you are in the testing site; not all locations will permit it. Once you get your test, on a small piece of paper write down everything you are afraid you will forget. It will take a minute or two but by dumping your worries onto the page you have effectively eliminated a certain amount of anxiety and driven off the panic you feel.

Get comfortable in your chair

Here is a clever technique that releases physical stress and helps you get comfortable, even relaxed in your body. You will tense and hold each of your muscles for just a few seconds. The trick is, you must tense them hard for the technique to work. You might want to practice this technique a few times at home; you do not want an unfamiliar technique to add to your stress just before a test, after all! Once you are at the test site, this exercise can always be done in the rest room or another quiet location.

Start with the muscles in your face then work down your body. Tense, squeeze and hold the muscles for a moment or two. Notice the feel of every muscle as you go down your body. Scowl to tense your forehead, pull in your chin to tense your neck. Squeeze your shoulders down to tense your back. Pull in your stomach all the way back to your ribs, make your lower back tight then stretch your fingers. Tense your leg muscles and calves then stretch your feet and your toes. You should be as stiff as a board throughout your entire body.

Now relax your muscles in reverse starting with your toes. Notice how all the muscles feel as you relax them one by one. Once you have released a muscle or set of muscles, allow them to remain relaxed as you proceed up your body. Focus on how you are feeling as all the tension leaves. Start breathing deeply when you get to your chest muscles. By the time you have found your chair, you will be so relaxed it will feel like bliss!

Fight distraction

A lucky few are able to focus deeply when taking an important examination, but most people are easily distracted, probably because they would rather be any place else! There are a number of things you can do to protect yourself from distraction.

Stay away from windows. If you select a seat near a window you may end gazing out at the landscape instead of paying attention to the work at hand. Furthermore, any sign of human activity, from a single individual walking by to a couple having an argument or exchanging a kiss will draw your attention away from your important work. What goes on outside should not be allowed to distract you.

Choose a seat away from the aisle so you do not become distracted by people who leave early. People who leave the exam room early are often the ones who fail. Do not compare your time to theirs.

Of course, you love your friends; that's why they are your friends! In the test room, however, they should become complete strangers inside your mind. Forget they are there. The first step is to physically distance yourself from friends or classmates. That way, you will not be tempted to glance at them to see how they are doing, and there will be no chance of eye contact that could either distract you or even lead to an accusation of cheating. Furthermore, if they are feeling stressed because they did not spend the focused time studying that you did, their anxiety is less likely to permeate your hard-earned calm.

Of course, you will want to choose a seat where there is sufficient light. Nothing is worse than trying to take an important examination under flickering lights or dim bulbs.

Ask the instructor or exam proctor to close the door if there is a lot of noise outside. If the instructor or proctor is unable to do so, block out the noise as best you can. Do not let anything disturb you.

Most tests do not allow any personal items in the exam room. Eat protein, complex carbohydrates and a little fat to keep you feeling full and to supercharge your energy. Nothing is worse than a sudden drop in blood sugar during an exam.

Do not allow yourself to become distracted by being too cold or hot. Regardless of the weather outside, carry a sweater, scarf or jacket if the air conditioning at the test site is set too high, or the heat set too low. By the same token, dress in layers so that you are prepared for a range of temperatures.

Watch Caffeine

Drinking a gallon of coffee or gulping a few energy drinks might seem like a great idea, but it is, in fact, a very bad one. Caffeine, pep pills or other artificial sources of energy are more likely to leave you feeling rushed and ragged. Your brain might be clicking along, all right, but chances are good it is not clicking along on the right track! Furthermore, drinking lots of coffee or energy drinks will mean frequent trips to the rest room. This will cut into the time you should be spending answering questions and is a distraction in itself, since each time you need to leave the room you lose focus. Pep pills will only make it harder for you to think straight when solving complicated problems on the exam.

At the same time, if anxiety is your problem try to find ways around using tranquilizers during test-taking time. Even medically prescribed anti-anxiety medication can make you less alert and even decrease your motivation. Being motivated is what you need to get you through an exam. If your anxiety is so bad that it threatens to interfere with your ability to take an exam, speak to your doctor and ask for documentation. Many testing sites will allow non-distracting test rooms, extended testing time and other accommodations as long as a doctor's note that explains the situation is made available.

Keep Breathing

It might not make a lot of sense, but when people become anxious, tense, or scared, their breathing becomes shallow and, in some cases, they stop breathing all together! Pay attention to your emotions, and when you are feeling worried, focus on your breathing. Take a moment to remind yourself to breathe deeply and regularly. Drawing in steady, deep breaths energizes the body. When you continue to breathe deeply you will notice you exhale all the tension.

It is a smart idea to rehearse breathing at home. With continued practice of this relaxation technique, you will begin to know the muscles that tense up under pressure. Call these your "signal muscles." These are the ones that will speak to you first, begging you to relax. Take the time to listen to those muscles and do as they ask. With just a little breathing practice, you will get into the habit of checking yourself regularly and when you realize you are tense, relaxation will become second nature.

Avoid Anxiety Before a Test

Manage your time effectively

This is a key to your success! You need blocks of uninterrupted time to study all the pertinent material. Creating and maintaining a schedule will help keep you on track, and will remind family members and friends that you are not available. Under no circumstances should you change your blocks of study time to accommodate someone else, or cancel a study session to do something more fun. Do not interfere with your study time for any reason!

Relax

Use whatever works best for you to relieve stress. Some folks like a good, calming stretch with yoga, others find expressing themselves through journaling to be useful. Some hit the floor for a series of crunches or planks, and still others take a slow stroll around the garden. Integrate a little relaxation time into your schedule, and treat that time, too, as sacred.

Eat healthy

Instead of reaching for the chips and chocolate, fresh fruits and vegetables are not only yummy but offer nutritional benefits that help relieve stress. Some foods accelerate stress instead of reducing it and should be avoided. Foods that add to higher anxiety include artificial sweeteners, candy and other sugary foods, carbonated sodas, chips, chocolate, eggs, fried foods, junk foods, processed foods, red meat, and other foods containing preservatives or heavy spices. Instead, eat a bowl of berries and some yogurt!

Get plenty of ZZZZZZZs

Do not cram or try to do an all-nighter. If you created a study schedule at the beginning, and if you have stuck with that schedule, have confidence! Staying up too late trying to cram in last-minute bits of information is going to leave you exhausted the next day. Besides, whatever new information you cram in will only displace all the important ideas you've spent weeks learning. Remember: You need to be alert and fully functional the day of the exam

Have confidence in yourself!

Everyone experiences some anxiety when taking a test, but exhibiting a positive attitude banishes anxiety and fills you with the knowledge you really do know what you need to know. This is your opportunity to show how well prepared you are. Go for it!

Be sure to take everything you need

Do not chitchat with friends

Let your friends know ahead of time that it is not anything personal, but you are going to ignore them in the test room! You need to find a seat away from doors and windows, one that has good lighting, and get comfortable. If other students are worried their anxiety could be detrimental to you; of course, you do not have to tell your friends that. If you are afraid they will be offended, tell them you are protecting them from your anxiety!

Common Test-Taking Mistakes

Taking a test is not much fun at best. When you take a test and make a stupid mistake that negatively affects your grade, it is natural to be very upset, especially when it is something that could have been easily avoided. So what are some of the common mistakes that are made on tests?

Do not fail to put your name on the test

How could you possibly forget to put your name on a test? You would be amazed at how often that happens. Very often, tests without names are thrown out immediately, resulting in a failing grade.

Marking the wrong multiple-choice answer

It is important to work at a steady pace, but that does not mean bolting through the questions. Be sure the answer you are marking is the one you mean to. If the bubble you need to fill in or the answer you need to circle is 'C', do not allow yourself to get distracted and select 'B' instead.

Answering a question twice

Some multiple-choice test questions have two very similar answers. If you are in too much of a hurry, you might select them both. Remember that only one answer is correct, so if you choose more than one, you have automatically failed that question.

Mishandling a difficult question

We recommend skipping difficult questions and returning to them later, but beware! First, be certain that you do return to the question. Circling the entire passage or placing a large question mark beside it will help you spot it when you are reviewing your test. Secondly, if you are not careful to skip the question, you can mess yourself up badly. Imagine that a question is too difficult and you decide to save it for later. You read the next question, which you know the answer, and you fill in that answer. You continue to the end of the test then return to the difficult question only to discover you didn't actually skip it! Instead, you inserted the answer to the following question in the spot reserved for the harder one, thus throwing off the remainder of your test!

Incorrectly Transferring an answer from scratch paper

This can happen easily if you are trying to hurry! Double check any answer you have figured out on scratch paper, and make sure what you have written on the test itself is an exact match!

Thinking too much

Oftentimes, your first thought is your best thought. If you worry yourself into insecurity, your self-doubts can trick you into choosing an incorrect answer when your first impulse was the right one!

Conclusion

CONGRATULATIONS! You have made it this far because you have applied yourself diligently to practicing for the exam and no doubt improved your potential score considerably! Passing your up-coming exam is a huge step in a journey that might be challenging at times but will be many times more rewarding and fulfilling. That is why being prepared is so important.

Good Luck!

Register for Free Updates and More Practice Test Questions

Register your purchase at https://www.test-preparation.ca/register/ for fast and convenient access to updates, errata, free test tips and more practice test questions.

https://www.youtube.com/user/MrTestPreparation

Online Resources

How to Prepare for a Test - The Ultimate Guide

https://www.test-preparation.ca/prepare-test/

Learning Styles - The Complete Guide

https://www.test-preparation.ca/learning-style/

Test Anxiety Secrets!

https://www.test-preparation.ca/test-anxiety/

Time Management on a Test

https://www.test-preparation.ca/time-management/

Flash Cards - The Complete Guide

https://www.test-preparation.ca/flash-cards/

Test Preparation Video Series

https://www.test-preparation.ca/test-video/

How to Memorize - The Complete Guide

https://www.test-preparation.ca/memorize/

www.ingramcontent.com/pod-product-compliance
Lightning Source LLC
Chambersburg PA
CBHW072007070526
44583CB00015B/1366